GENERAL RELATIVITY:
A Guide to its Consequences for Gravity and Cosmology

GENERAL RELATIVITY:
A Guide to its Consequences for Gravity and Cosmology

J. L. MARTIN, B.Sc., B.A., Ph.D.
Reader in Physics
King's College (KQC), University of London

ELLIS HORWOOD LIMITED
Publishers · Chichester

Halsted Press: a division of
JOHN WILEY & SONS
New York · Chichester · Brisbane · Toronto

First published in 1988 by
ELLIS HORWOOD LIMITED
Market Cross House, Cooper Street,
Chichester, West Sussex, PO19 1EB, England
The publisher's colophon is reproduced from James Gillison's drawing of the ancient Market Cross, Chichester.

Distributors:
Australia and New Zealand:
JACARANDA WILEY LIMITED
GPO Box 859, Brisbane, Queensland 4001, Australia
Canada:
JOHN WILEY & SONS CANADA LIMITED
22 Worcester Road, Rexdale, Ontario, Canada
Europe and Africa:
JOHN WILEY & SONS LIMITED
Baffins Lane, Chichester, West Sussex, England
North and South America and the rest of the world:
Halsted Press: a division of
JOHN WILEY & SONS
605 Third Avenue, New York, NY 10158, USA
South-East Asia
JOHN WILEY & SONS (SEA) PTE LIMITED
37 Jalan Pemimpin # 05–04
Block B, Union Industrial Building, Singapore 2057
Indian Subcontinent
WILEY EASTERN LIMITED
4835/24 Ansari Road
Daryaganj, New Delhi 110002, India

British Library Cataloguing in Publication Data
Martin, J. L. (John Legot), *1931–*
General relativity: a guide to the consequences for gravity and cosmology.
1. Physics. General theory of relativity
I. Title
530.1′1

Library of Congress Card No. 88–21957

ISBN 0–7458–0519–1 (Ellis Horwood Limited)
ISBN 0–470–21183–0 (Halsted Press)

Printed in Great Britain by Hartnolls, Bodmin

Contents

For Margaret, who has been very patient.

Star death

Birth'd they were and cradled on the height
of furious Creation's blazing start.
Held up by raging chaos at their heart
and dancing on the radiant outflow bright.

Not for ever can they win this fight.
The dying fires will cease to act their part.
The pressure fails. No longer held apart
their day is over, giving place to night.

There lies a singularity in wait.
Past the horizon (which to pass is death)
with world lines jostling, fighting for each breath,
racked by the tide, they plunge toward their fate.
Nor is there any time to question how
the rest of space will fare: their end is now.

Preface

Nature and Nature's laws lay hid in night:
God said 'Let Newton be!' and all was light.

—Alexander Pope, *Epitaph intended for Sir Isaac Newton.*

It did not last: the Devil howling, 'Ho!
Let Einstein be!' restored the status quo.

—Sir John Squire.

I have little patience with scientists who take a board of wood, look for its thinnest part, and drill a great number of holes where drilling is easy.

—Albert Einstein.

There are at least three barriers in the way to a full understanding of Einstein's General Relativity. The first is the inadequacy of our imagination. We know what a curved surface looks like; we are utterly unable to imagine a curved three-dimensional space, let alone the 'curved' four-dimensional spacetime which is the subject of Einstein's great work. It doesn't much help to remember that the word 'curvature' is actually a metaphor—the mathematics of Einstein spacetime is *isomorphic* to the mathematics of curvature developed in the nineteenth century; our prejudices still lead us to think of perfectly reasonable mathematics and perfectly reasonable physics as being in some way startling or outrageous—*startling*, if we approach the subject with an open mind; *outrageous* to those people, and still there are many, whose preconceptions utterly prevent them from believing that any of it can possibly be true. The answer to this difficulty is to allow ourselves to be driven by the logic of the mathematics.

The second barrier is the scale of the algebra. In general terms, the mathematical apparatus needed to deal with curvature is fairly simple. Indeed, I believe that in Cambridge in the early thirties it was suggested that differential geometry ought not to be included in the undergraduate course: it had become too much of a soft option. I agree, as long as we talk in general terms. But the physicist is interested in actual observations on individual systems, and it is when we wish to move from the general to the particular that the algebra seems to expand without limit. I am not happy with the books which use formulas like 'after some tedious algebra'/'an exercise for the student'/etc., and I have gone to some trouble in explaining ways of coping with the work reliably. These days, serious workers in the field use electronic computers when the going gets heavy.

Finally, Einstein's field equations for gravity are appallingly nonlinear, and very few exact solutions are known. Fortunately, 'everyday' gravity is extremely weak, and excellent approximations can be made. But some of the most interesting aspects of the theory concern *very strong* gravity—gravitational collapse, black holes, close binary systems of neutron stars, cosmology—and any exact treatment at the level of this book can touch only the simplest cases.

There may be a fourth barrier. General Relativity may be viewed from two standpoints. A mathematician may be interested in the theory primarily as an example of the structures of differential geometry. A physicist will want to apply the theory to his observed environment. It is confusing to mix these aims; in particular, the reader may too easily become convinced that the perfectly proper jargon of the mathematician is in some way essential to an understanding of the physics. A real definition from another book on General Relativity goes: The *tangent vector* to a curve at the point P is the equivalence class of all curves tangential to the curve at P. Well, yes, so it is, and perfectly proper in its place. But not here; a physicist will probably prefer a less esoteric definition of a tangent. This is a book for physicists who want to know how the theory connects with observation, and the enticing byways of irrelevant mathematics will have to be left to other books. (On the other hand, the *relevant* mathematics is all here.) One practical effect of this policy is that the *components* of vectors and tensors are always to the fore, since real calculations are carried out on the components.

Tidal forces are given more attention than is usual. They are worth the trouble, since they are the manifestation of *genuine* gravity, rather than the pseudogravity arising from an unacknowledged acceleration. (One may also suspect that tidal forces need more attention from certain writers of science fiction; see Problem 8 of Chapter 9.)

I have made some assumptions about the knowledge of the reader. Anyone unfamiliar with at least elementary Newtonian mechanics and Newtonian gravitation is unlikely to make much headway, if only because they will be unable to see the point of it all. A reasonable knowledge of calculus up to at least the solution of simple differential equations is essential. The sum, product, and inverse of square matrices are used, as is the concept of the eigenvalues of a square matrix. A reasonable feel for geometry in three dimensions will be a help. The only other prerequisite is a stout-hearted dedication.

This book is an expansion of a third-year undergraduate course given in the Department of Physics, King's College London. I have always been grateful for the enthusiasm of the students for the subject; it has made the teaching so much more of a pleasure. In addition, it has surprised and encouraged me that, in an age where science is too often regarded as important only for technology or profit, there should be so many people who want to include such a subject in their preparation to meet a competitive world outside. Science for Science' sake is not yet dead.

JLM

King's College London
March 1988

1

The principle of equivalence

1.1 The relativity of Galileo and Newton

It is fair to say that mechanics was not at all understood before the seventeenth century. Nowadays we tend to take it for granted that force produces acceleration, but the ancients were sure that a force is needed to generate *velocity*: after all, if the horse stops, so does the cart. All kinds of bizarre explanations were given for difficult questions like: why does an arrow continue to move after it has left the bow? The Universe has a natural direction, namely *down*, in which earthly bodies are inclined to fall. Heavenly bodies stay up by being fixed to transparent spheres with complicated motions. The Earth was believed by many to be flat (some notions seem to persist after their time). And so on.

Against this background, Isaac Newton's First Law was a triumph: an undisturbed body moves in a straight line at constant speed. To grasp this truth in an environment where there are no undisturbed bodies and no constant speeds, and where no trajectory is a straight line, indicates the genius of Newton, and of his immediate predecessors like Galileo. To alter a velocity needs a force (in a way, this is the definition of force), and the other laws of motion take care of that aspect. Experimentally, the best tests of the new theories were provided by celestial mechanics, where dissipative forces are tiny, and the observations of the astronomers were made with unparalleled precision. Johannes Kepler in particular had summed up the motions of the planets in a few empirical laws (one of which we would now recognize as the conservation of angular momentum), and Newton was able to show that these laws were fully consistent with an inverse-square law of gravitational force.

One of the consequences of the new mechanics was a kind of *relativity*: no observer is able to determine his own velocity in any absolute sense; his velocity is always relative to something or someone else. These days we are familiar with this; in a smoothly flying aircraft we cannot tell our speed without looking out of the window, and we can cope with our coffee in the air as competently as we can on the ground.

The ground itself, of course, is moving at no inconsiderable speed as the Earth rotates. (All of this was understood by Newton, so it is amusing to recall that two centuries later the new railways were criticized because the passengers would be crushed to death by the excessive speed of the trains.)

1.2 The Michelson–Morley experiment

Old ideas die hard, and even though there was no theoretical need and no observational evidence for absolute velocity or, equivalently, for a state of absolute rest, physicists still clung to it. This was particularly true with the arrival of Maxwell's equations in the nineteenth century: it was clear that these equations included wave motions among their solutions. If there was a wave, surely there must be something to do the waving, and the notion of an all-pervading **luminiferous aether** was born.

In 1881, Albert Michelson carried out an experiment (later improved with Morley) to find out how fast his laboratory was moving through the aether. The principle was that the length of an optical path in an interferometer ought to depend on the orientation of the path in relation to the aether flow (see Problem 1, at the end of this chapter). When the experiment was first carried out, no aether flow was detected. Of course, this could have meant that the laboratory was momentarily and accidentally at rest with respect to the aether; the solution was to wait six months, when the Earth was at the other side of its orbit with an easily observable velocity of—presumably—as much as $60\,\mathrm{km\,s^{-1}}$. Still no velocity was found. Over the years, improved techniques have led to an upper limit on the speed of aether flow of about $5\,\mathrm{cm\,s^{-1}}$; and it is generally accepted that there is no aether flow at all.

All attempts to detect the presence of an aether have failed. From the physicist's standpoint, what cannot be detected isn't there, and we need never mention it again. Additionally,

> **the velocity of light is to be the same, in whatever direction, for any observer whatever, no matter what the relative motion of two observers may be.**

Further, all velocities are relative.

This result is entirely in the spirit of Newton and Galileo. This is worth remembering as we move on to consequences that our intuition may be unhappy with at first.

1.3 Kinematic forces

Imagine that we work in a laboratory without windows, mounted on a rotating turntable. We will be aware of a field of force directed from the centre of the room towards the walls. As we walk around, we shall be aware of a transverse force making it difficult to walk in a straight line. These are of course the familiar centrifugal and Coriolis forces. It is often asked: are they real? Of course they are: they can make life uncomfortable for the denizens of the room and, more to the point, they can be measured with accelerometers (masses linked with springs are simplest). Why they are often supposed to be fictitious is because we know very well how they can be removed, or at least explained; in terms of a more appropriate frame of reference they are seen to be not so much forces as accelerations consequent on the rotation. We shall call forces which are 'really' accelerations **kinematic forces**.

How is a kinematic force recognized? It has the important property, not shared with other forces (electrostatic, for example):

a kinematic force is strictly proportional to the mass on which it acts.

This book is almost entirely about kinematic forces.

1.4 *mg*?

The exciting example of a kinematic force is gravity. Ever since Galileo is supposed to have dropped objects from the Leaning Tower in Pisa, it has been understood that everything falls with the same acceleration g: gravity is 'really' acceleration in disguise. (A common way of putting this is: Inertial and gravitational mass are equal.)

This is such an important matter that a great deal of experimental effort has been put into its verification. The technique is to compare gravity with some other force which is *known* to be kinematic. The favourite is the centrifugal force arising from the Earth's rotation; this is a smallish, but not insignificant, force. The early experiments were done by the Baron Eötvös in 1891. Equal masses of very dissimilar materials were hung from the ends of the beam of a torsion balance oriented east–west. Rotating the case of the balance through 180 degrees will not rotate the equilibrium position of the balance through the same angle, *unless* gravity is a kinematic force. Eötvös was unable to find any change in the position of equilibrium. (See Problem 4.)

In 1962, R.H. Dicke did experiments of the same kind to compare the centrifugal force on the Earth with the gravitational force of the Sun. Rotating the case of the balance relative to the Sun is simpler: wait for twelve hours, and the Earth will do it for you. As a result, much steadier mountings for the experiment are possible, and the improvement in this and in subsequent experiments shows that gravity is kinematic to within one part in 10^{11}.

1.5 Inertial observers

Kinematic forces can always be transformed away. An experimental procedure is (i) to turn off all nonkinematic forces and (ii) to prevent rotation. For example, as I stand on the floor of a room, I experience two forces: one is gravity, and the other is the 'real' reaction of the floor on my feet. Turn off the second of these (remove the floor), and I shall immediately become 'weightless' (my accelerometer will register zero). How long I shall remain so will depend on the height of the room below; as soon as I meet the floor, gravity will reappear. For the time being, however, I am an inertial observer. Of course, a stationary bystander will insist that gravity is there all the time, responsible for the acceleration of my fall. Who is right? Is it not evident that this is exactly the kind of question we like to ask about centrifugal force?

An **inertial observer** is an observer in free fall and not rotating. It is important that an observer can tell whether he is inertial or not, without needing to refer in any way to the rest of the Universe. His laboratory must be supplied with enough accelerometers to measure gravitational and centrifugal forces adequately. If all the accelerometers show zero, then he is inertial; if they don't, then he isn't. There is a sharp contrast here: an observer in his windowless laboratory cannot measure his own velocity, but he can measure his own acceleration without ambiguity: acceleration is absolute; velocity is relative.

This concept of an *inertial observer* will be crucial from now on.

1.6 The Principle of Equivalence

It would seem likely that one inertial observer will be much like another. This idea is at the heart of Einstein relativity, and it has been given the status of a Principle.

> **The Principle of Equivalence: We suppose two isolated inertial laboratories. Then all the laws of physics in either laboratory are precisely the same as the laws in the other.**

(There is a tiny proviso to be considered later.)

In view of the Michelson–Morley experiment, we might reasonably expect such a Principle, since an obvious instance would be the constancy and isotropy of the velocity of light. However, it covers far more than that: it is very strong indeed. It implies, for example, that the laws of physics for an inertial observer in the first minute, say, after the creation of the Universe are not different from our own. This is still a matter for discussion, but we shall not consider it further here: the Principle will be taken as true.

The original version of the Principle (nowadays known as the *weak* Principle) said simply that gravity and acceleration are equivalent or—to use the language introduced above—gravity is a kinematic force.

The full consequences of the Principle need the techniques of Special and General Relativity for their precise elaboration. However, certain situations can be handled in a very simple approximate way, and we shall now look at a few which provide a good flavour of the kinds of surprise that lie in wait.

1.7 Light is heavy!

An experimenter in his inertial laboratory transmits a pulse of light from the centre of one wall to the centre of the opposite wall. Of course, this pulse is seen to travel in a straight line with speed c.

It so happens that the laboratory is in free fall towards the Earth, floor first. A bystander on the Earth's surface watches events from his non-inertial point of view. For consistency, he must see exactly the same course of events. Thus he too sees the pulse move along the same straight line from wall to wall; the difference is that for him the straight line itself is falling towards the Earth with an acceleration g. For him, the trajectory of the pulse is a parabola of free fall towards the Earth: the Earth attracts the pulse as it would any particle.

That light should fall towards a massive body is not a new idea. In 1801, for example, J. Soldner calculated that a star viewed near the Sun should appear to be displaced through an angle 0.85 seconds of arc. This was at a time when it was still widely believed that light is constituted of particles, not waves; the century-long controversy had still not been settled. If light were particles, it would be easy to imagine that it would be deflected by the gravity of the Sun, just like any comet or other object. Soldner's result was echoed in 1911 by Einstein, who used the (weak)

Principle of Equivalence to argue that electromagnetic energy ought to gravitate, and therefore that a deflection should be observed. The more careful treatment of full General Relativity increases Soldner's result by a factor of two.

In 1919, Arthur Eddington observed the predicted deflection of the light from a star as the ray grazed the edge of the Sun during a total eclipse. The procedure was difficult, mainly on account of the abrupt temperature change as a solar eclipse approaches totality; this generates massive atmospheric turbulence, along with considerable thermal distortions in the apparatus itself. More recent optical experiments of this kind with refined apparatus have done little to improve accuracy. The most reliable modern verifications of the effect use observations of radio sources as they are occulted by the Sun; these are far less at the mercy of atmospheric conditions.

1.8 The gravitational redshift

The same experimenter in the same inertial laboratory transmits a pulse of light from the centre of the floor to the centre of the ceiling. Again, the pulse is seen to travel in a straight line with speed c and, of course, its frequency ν_0 does not change. What does the Earth-bound bystander see now?

To begin with, if the laboratory is falling with velocity v at the moment of transmission, he will see a Doppler-shifted frequency in the normal way; to sufficient accuracy,

$$\nu_1 = \nu_0(1 - v/c).$$

There is no surprise here. But the pulse takes a time H/c to reach the ceiling (at a height H above the floor), and during this time the laboratory has accelerated somewhat, to a velocity $v + gH/c$. At the moment of receipt, the bystander sees a *different* Doppler-shifted frequency

$$\nu_2 = \nu_0(1 - [v + gH/c]/c);$$

in rising through a vertical distance H, the pulse has changed frequency by an amount given by

$$\nu_2 = \nu_1(1 - gHc^{-2}).$$

The frequency has decreased; if the inertial experimenter sees no change in the frequency, then the non-inertial bystander must see a **redshift**. Since the amount of the shift depends on g, it must find its origin in the gravity field of the Earth.

The gravitational redshift was first observed with any certainty by Pound and Rebka in 1960, using a vertical separation of 23 m. The change over such a small altitude difference is tiny, and detecting it required exotic techniques. Certain nuclear transitions emit gamma-photons with extraordinarily narrow line-width, so narrow that the redshift of such a photon may render it incapable of stimulating the reverse transition at the higher altitude, unless steps are taken to shift the resonance by a measured amount. The arrival of transportable, yet reliable, atomic clocks has made the effect easier to see; a clock in a high-flying aircraft is compared by radio with a similar clock on the ground. Using specially designed equipment, Brault has been able to observe the shift in the visible spectrum of the Sun.

1.9 Mass and energy are the same

In the last section, gH is the change in Newtonian gravitational potential $\phi(z)$ from floor to ceiling. Thus we may rearrange the last equation approximately to get

$$\nu_1 + \nu_1\phi(z_1)/c^2 = \nu_2 + \nu_2\phi(z_2)/c^2.$$

Writing $E = h\nu$, where h is the Planck constant, we have

$$E + (E/c^2)\phi(z) \quad \text{does not change.}$$

This is nothing else than conservation of energy. The first term is the energy of a photon in the pulse. The second term is the potential energy of a particle of *mass* E/c^2 in the gravitational potential $\phi(z)$. This is an example of the celebrated $E = Mc^2$. Mass and energy are the same thing, expressed in different units.

The equivalence of mass and energy is a cornerstone of radioactive decay; to be unstable, a nucleus must be more massive than the collection of its decay products—the excess mass goes into kinetic energy.

1.10 Moving clocks run slow

Now imagine a non-inertial laboratory, in free fall, but rotating about an axis from floor to ceiling. Our experimenter will be aware of a force field acting from the centre of the room towards the walls; the magnitude of the force is $r\omega^2$ per unit mass, where r is the distance from the axis and ω is the (constant) angular velocity of the rotation. Being good at theory, he writes down the potential for this force,

$$\phi(r) = -\tfrac{1}{2}r^2\omega^2,$$

and claims that, for a photon moving out from the axis,

$$\nu + \nu\phi(r)/c^2 \quad \text{does not change,}$$

and that therefore

$$\nu(r) = \nu(0)/(1 - r^2\omega^2/2c^2).$$

As the photon moves away from the axis its frequency increases.

What is an *inertial* bystander to make of all this? For him the photon's frequency has to be constant throughout its flight. However, he can see the rotation for what it is; in particular, he can see that the clock that the experimenter might use to measure the frequency at a distance r from the axis is moving with a transverse velocity $v = r\omega$. Agreement will be restored if such a clock runs slow by an appropriate amount. In fact, the clock rate will need to be reduced by a factor $1 - v^2/2c^2$. Moving clocks run slow.

The rotating laboratory experiment was done in 1961 by Hay, Schiffer, Cranshaw, and Egelstaff. The techniques were similar to those of Pound and Rebka; a photon emitted at the centre of a rotating disc was found to be out of resonance at the moving periphery, by the predicted amount.

Twenty years before, it was observed that cosmic-ray mu-mesons, generated in the upper atmosphere, arrive at the surface of the Earth in much greater numbers than might be expected, given their velocity and laboratory lifetime; they ought to have decayed on the way. Harmony is restored if we suppose that the decay 'runs slow', just as if it were a moving clock. The lengthening of lifetime with velocity has been fully vindicated since.

We could continue. Systematic use of the Principle of Equivalence leads to many remarkable results which are fully in accord with experiment. All in all, available evidence unanimously supports the Principle, and it is now regarded by most physicists as unshakeable.

1.11 What then *is* gravity?

In the middle of the nineteenth century, Riemann generalized the work of Gauss and others on curved surfaces and 3D space to spaces of any number of dimensions. In so doing, he provided the techniques which are necessary for General Relativity, and we shall see a great deal of them later. He suspected that gravity is by nature a kinematic force, and it occurred to him to ask whether the curved trajectories of objects in free fall might be explained by an underlying *curvature* of 3D space. After all, a curvature of the environment would be expected to influence all objects in an identical manner, whatever their structure. The idea was unworkable, and was soon dropped.

At the turn of the century it was recognized that the productive way forward was to weld the points of 3D space and the instants of 1D time into a $(1 + 3)$D structure—**spacetime**—of *events*. If left uncurved, this spacetime provided the ideal backcloth for phenomena far from any gravitational influences, and is the foundation of the first theory, now called Special Relativity. This theory copes with the Michelson–Morley result on the constancy of the velocity of light, and allows for all kinematic forces *except* gravity. If curvature is let in, so is gravity. We then have Einstein's General Relativity.

1.12 Why haven't we noticed?

If our environment is curved enough to produce the very obvious effects of gravity, why is it not obvious in other ways? Surely astronomers, who are surveyors of space on the grand scale, would have noticed long ago.

On account of the speed of light, which is huge by ordinary standards, even the slightest of curvatures can produce substantial effects. For example, what kind of radius of curvature would we need to generate our familiar terrestrial acceleration $g = 10\,\mathrm{m\,s^{-2}}$? We shall have to combine g with $c = 3 \times 10^8\,\mathrm{m\,s^{-1}}$ to obtain something with the dimensions of *length*; the only possibility is $c^2/g = 9 \times 10^{15}\,\mathrm{m}$. This gives an estimate of the required radius of curvature, and it is enormous. The curvature, therefore, is very slight indeed. The Earth does not distort the environment very much.

How should we estimate the mass of the Earth, from this point of view? The acceleration generated by the presence of the Earth obeys the inverse-square law: $g(r) = \text{constant}/r^2$. Using c to transform units again, we write

$$g(r)/c^2 = m/r^2$$

where m is a *length* which describes the gravitational effect of the Earth. From known values at the surface of the Earth ($r = 6378$ km), we evaluate

$$m = 0.45 \text{ cm}.$$

This is the **mass of the Earth**, in unfamiliar units. It is tiny in comparison with the scale of the setting.

It is really not surprising that the curvature went unnoticed for so long.

1.13 The small print

Now for the small proviso mentioned earlier. Go back to the experimenter in his laboratory free-falling towards the Earth. Objects near the floor are not quite 'weightless': they will drift 'down' towards the floor. Objects near the ceiling will drift 'up'. Objects near the walls will drift inwards. This is because the gravitational field of the Earth is not homogeneous; it varies in strength and direction from place to place, and what the experimenter experiences are the residual inhomogeneities after the major acceleration has been subtracted. These are the **tidal forces**. (The ocean tides on Earth are the result of the tidal forces of the Moon and, to a lesser extent, the Sun.) They are kinematic forces, *but they can never be abolished completely in a laboratory of finite size*. The Principle of Equivalence must always be formulated with this in mind.

This gives the clue to a conundrum which puzzles many. An object in an inertial laboratory is electrically charged; does it radiate? The usual answer is: of course it doesn't. The usual answer is wrong. For think of this: a charged object held stationary at the surface of the Earth certainly does not radiate. If it is allowed to drop, it will begin to radiate in the manner of all electric charges undergoing acceleration, even though it is now in an inertial environment. Why?

The answer is in the tidal forces, which are changing as the object falls. These interact in a time-dependent manner with the inverse-square electric field of the charge, which extends out far enough from the charge itself to sense the inhomogeneities. The result is that radiation is generated. The details of radiation from an accelerated charge are difficult, and far beyond the scope of this book.

Readers with their wits about them will remark that a massive object carries with it an inverse-square *gravity* field at all times: perhaps this ought to generate gravity radiation in appropriate circumstances, for the same kind of reason. Indeed it does, if the predictions of General Relativity are correct. Gravity radiation is considered at a basic level in Chapter 13.

1—Notes and problems

1. Michelson's interferometer uses a partially reflecting mirror M to split an incoming light beam into two. After reflection at two mirrors A and B, the beams are reunited at M (Fig. 1.1). The lengths of the two arms of the interferometer are equal (L). Differences in the times of traverse in the two arms may be detected by a shift in the interference fringe pattern seen at M. For simplicity, we consider the case where one arm lies parallel to V, the velocity of the interferometer through the aether.

   Work in the laboratory frame, and show that the laboratory velocities of light

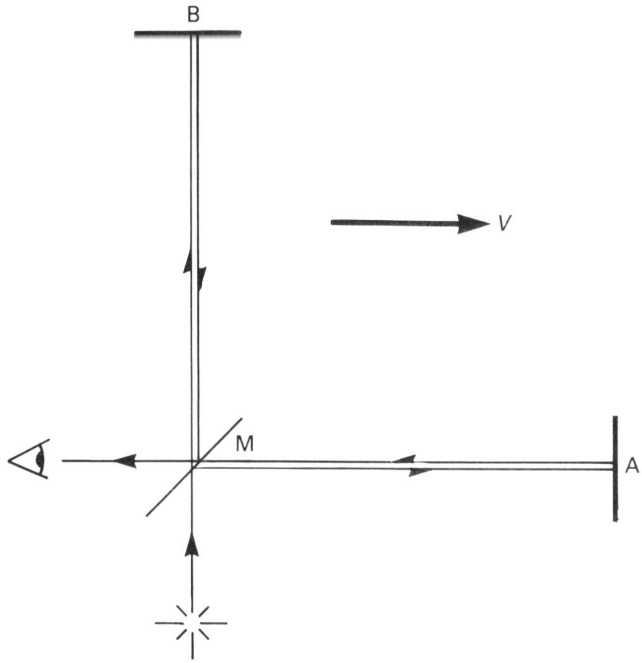

Fig. 1.1

in the arm MAM are $c - V$ one way and $c + V$ the other, while the velocity in MBM is $(c^2 - V^2)^{1/2}$ each way. Hence show that the time taken to traverse MAM is $2L\gamma^2/c$, while the different time for MBM is $2L\gamma/c$. Here $\gamma = (1 - V^2/c^2)^{-1/2}$.

The experimental null result implies, of course, that the discrepancy γ is absent, showing that there is something deeply wrong with the above discussion.

2. In the early days, before Special Relativity, the **Fitzgerald–Lorentz contraction** was proposed as a way of making sense of the null result of the Michelson–Morley experiment: all material objects were supposed to be *contracted* by a factor γ in their direction of motion. Transverse dimensions were unaffected. This factor was chosen to ensure that the different path-lengths in the interferometer were the same.

   Show that the Fitzgerald–Lorentz contraction ensures a null result for *all* orientations of the interferometer, and not only for those where one arm is parallel to the aether flow.

   The idea is not as bizarre as it sounds. The molecular structure of the interferometer is controlled by electromagnetic forces, governed by exactly the same laws—Maxwell's equations—as govern the propagation of light. It is therefore not difficult to imagine a conspiracy wherein a change of light velocity due to an aether flow is precisely compensated for by an electromagnetic shrinkage due to the same aether flow. This would be enough to make the aether flow unobservable *in principle*. We shall leave it to the philosophers to discuss in what sense the aether may be said to exist in such circumstances.

3. A 100 m-high mast is erected at the Earth's equator, and is perfectly vertical

according to plumbline measurements. A visitor at the top of the mast drops an object from rest. Explain why he concludes that the mast is out of true by about 3.25 cm.

4. If gravity is not a kinematic force, then a plumbline vertical (except at the Poles or at the Equator) will depend on the composition of the bob. Explain.
   In essence, the Eötvös experiment consists in hanging different plumblines at the ends of the beam of a sensitive torsion balance. Any discrepancy in the apparent verticals results in a torque whose sense reverses when the balance is turned through 180 degrees.

5. The essence of the technique used by Pound, Rebka and Snider—and by Hay and his colleagues—depends on a certain type of nuclear reaction; an example is

$$\begin{aligned} {}^{57}\mathrm{Co} + \mathrm{e} &\rightarrow {}^{57}\mathrm{Fe}\,(\text{excited}) \qquad (\text{by } K\text{-capture}) \\ &\rightarrow {}^{57}\mathrm{Fe}\,(\text{ground}) + \text{photon}. \end{aligned}$$

   With care, it is possible to confine the frequency of the emitted photon within extremely narrow limits. In fact, the emitter is anchored within a crystal structure of macroscopic size, thus making its recoil—and therefore Doppler consequences—utterly negligible in a substantial proportion of events (as a result of a quantum phenomenon known as the **Mössbauer effect**). The photon travels to another part of the laboratory, where it is given the chance to be absorbed by resonance, according to

$${}^{57}\mathrm{Fe}\,(\text{ground}) + \text{photon} \rightarrow {}^{57}\mathrm{Fe}\,(\text{excited}).$$

   Again, with use of the Mössbauer effect, the resonance absorption can be kept within narrow limits. The profile of the absorption is readily obtained with the help of the Doppler effect: moving either the emitter or the absorber at very small speeds is quite sufficient.
   At what speed, and in what sense, must the emitter be moved to restore resonance? (Assume a vertical separation of 23 m. Hint: relate v/c to gH/c^2.)

6. A spaceship in empty space undergoes a steady acceleration g. Give reasons for supposing that a clock at the stern goes slow in comparison with a similar clock at the nose. (Consider the same ship at rest on its launching pad, and use the Principle of Equivalence.)

7. An aircraft flies at speed V at constant altitude H. Show that—with neglect of the rotation of the Earth—if $2gH = V^2$ a clock on board will keep good time with a similar clock on the ground vertically below. Are 'reasonable' values of H and V possible?

8. Two synchronized atomic clocks are each flown once round the Equator of the Earth, in opposite directions. Which clock now shows the later time?

9. An Earth satellite travels with speed V in a circular orbit of radius r; its radial acceleration is V^2/r. Show that $rV^2 = mc^2$, where m is the 'mass' of the Earth, expressed as a length.
   To stay in circular orbit at an altitude of 250 km, an Earth satellite requires a speed of $7759\ \mathrm{m\,s^{-1}}$. Conclude that $m = 4.47$ mm. (Radius of Earth: 6378 km. Speed of light: $2.998 \times 10^8\ \mathrm{m\,s^{-1}}$.)

The masses of terrestrial objects may be very accurately known in kilograms, and poorly known in metres. The reverse is true for large astronomical objects, whose masses are determined, not by 'weighing', but by the motions of nearby probes.

10. Repeat the calculation, using the Moon as the probe satellite. (Earth–Moon distance: 3.844×10^8 m. Period of orbit: 27.32 days.) The reason for the slight overestimate for m in this case is the relatively large mass of the Moon, about 1/80 of the mass of the Earth, which needs to be taken into account.

11. In a similar way, show that the mass of the Sun is 1.487 km. (Sun–Earth distance: 1.496×10^{11} m. The period of the orbit is well known.)

2

Special relativity

There are many ways to introduce the formalism of Special Relativity. Some are much more abstract than others. The route we shall take is due to Hermann Bondi, and has the advantage of starting from concrete observations of a very familiar kind: **radar** is used to map out the environment of the observer in an essentially obvious way. (Radar—**ra**dio **d**etection **a**nd **r**anging—was invented in the 1940s for military purposes. We shall find the underlying concepts very useful.)

Our observers will always possess a radar set and a full set of accelerometers. A radar set comprises a transmitter of short radio pulses, a receiver to detect the return of those pulses echoed from the environment, and a clock to record the times of transmission and receipt. The accelerometers are there to ensure that the observer is *inertial* at all times, at least in the early parts of the discussion.

2.1 The radar formulas

An inertial observer—whom we shall call A—investigates his environment by transmitting radio pulses and observing the reflected echoes. First, he has measured the velocity of such pulses in his own laboratory, and has found it to have magnitude c and to be independent of direction. Now he sends out a pulse at time T_1 and observes, at the later time T_2, an echo of that pulse from an object P. Conventionally, but very naturally, A calculates the **distance** r of the object P, determined at the **time** t, by the **radar formulas**

$$r = c(T_2 - T_1)/2 \quad \text{and} \quad t = (T_2 + T_1)/2. \tag{2.1}$$

The act of P reflecting the pulse is an example of an **event**. The observer A assigns to this event the **coordinates** t and r (and additionally, if he happens to be in three dimensions, two further coordinates to specify the direction of transmission and receipt: the usual angles θ and ϕ of polar coordinates will do). To label the set of all possible events in 3D space requires *four* coordinates, since the time of an event needs to be specified in addition to its position: the events are the 'points' in a manifold of *four* dimensions, which we shall refer to as (1 + 3)D **spacetime**.

2.2 The Minkowski diagram

The **Minkowski diagram** is, for the moment, a plot of the events observed by A. Being 4D, it is difficult to draw, and we have to content ourselves with partial representations of the diagram; nevertheless, even partial diagrams are a real help. For example, consider only those objects which lie at all times in a constant direction from A, that is, which move only radially towards or away from A, along the x-axis, say. The course of events is then straightforwardly plotted in an xt-plot; an example is shown in Fig. 2.1. The motion of the object P is represented by the **worldline** of P. The worldlines of radar pulses are straight lines $x \pm ct = \text{constant}$, and the worldline of the observer A is the t-axis ($x = 0$).

2.3 Uniform velocity and the Bondi *K*-factor

If the worldline of P is a straight line in the Minkowski diagram of A, then we say that 'P moves with **uniform velocity**' with respect to A. If P moves in the x-direction

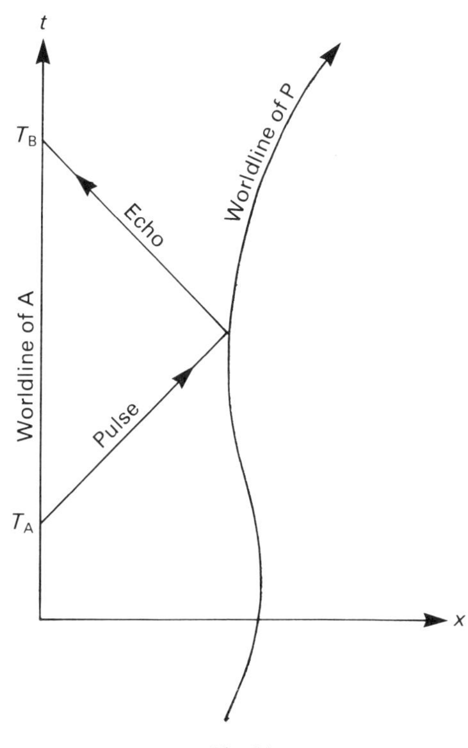

Fig. 2.1

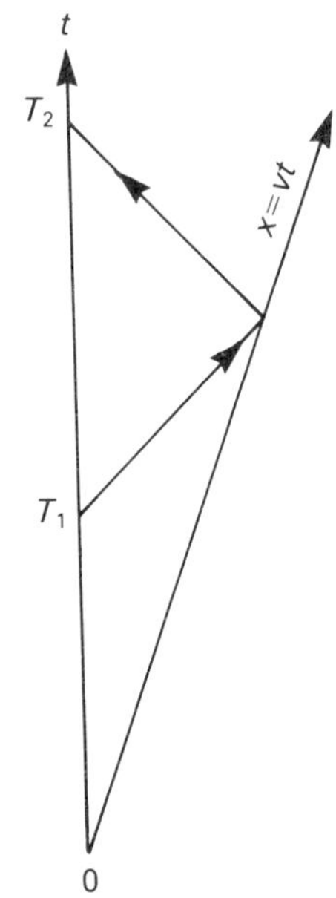

Fig. 2.2

with speed v and, for convenience, leaves A at $t = 0$, then the equation of the worldline is

$$x = vt. \tag{2.2}$$

This clearly accords with what we usually mean by speed.

The object P is observed by radar (with observed times T_1 and T_2) (Fig. 2.2). Then substituting the radar formulas in (2.2) gives

$$c(T_2 - T_1)/2 = v(T_2 + T_1)/2,$$

which leads to the useful relation

$$T_2 = T_1(c + v)/(c - v) = T_1 K^2. \tag{2.3}$$

Here the **Bondi *K*-factor** is defined by

$$K^2 = (c + v)/(c - v). \tag{2.4}$$

2.4 Special Relativity

Let us now imagine *two* inertial observers. Each has determined privately that his own instruments are showing zero acceleration. How should these observers regard each other? It is tempting to assert that the observers must be moving with uniform relative velocity. This is *not at all obvious*, since it is in fact not always true in experience! Two Earth satellites in different orbits are each inertial (each being in 'free fall'), yet their relative velocity is not uniform. Clearly, keeping away from all sources of gravity is a necessary condition for a pair of inertial observers to have uniform relative velocity. We accept this restriction for the moment, but shall need to see how to relax it later.

So: the fundamental ideas of Special Relativity may be expressed in terms of the mutual relation of a pair of inertial observers. If A and B are two observers, each inertial, then

(i) B moves with uniform relative velocity with respect to A, and vice versa;
(ii) Apart from sign, the relative velocities are equal.

So far, there is nothing that is not true also for Newtonian mechanics. The crucial departure comes when we remember the Principle of Equivalence, as evidenced by the Michelson–Morley observations on the velocity of light; we need to add

(iii) For both A and B, the velocity of light is the same in all directions, and has the same magnitude c.

The symmetry between A and B is perfect.

In the light of this, consider as before the case of A and B separating radially with relative speed v. Imagine a single radar pulse bounced to and fro between A and B

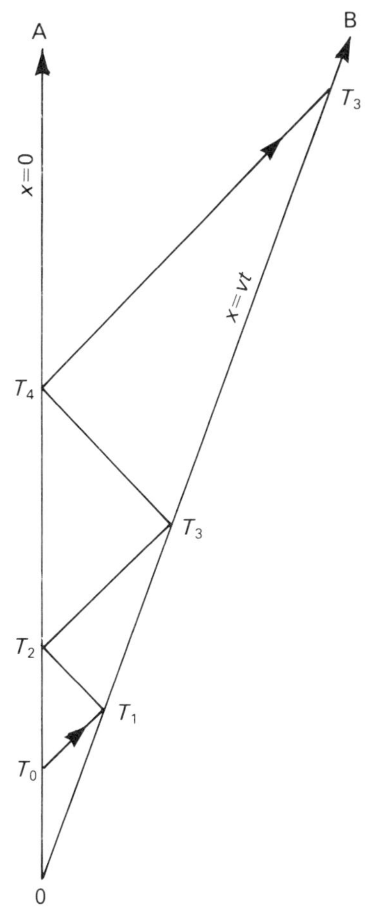

Fig. 2.3

(Fig. 2.3). A observes the pulse at times (on his clock) $T_0, T_2, \ldots$, while B observes it at times (on his own similar clock) $T_1, T_3, \ldots$.

On account of the symmetry between A and B, their K-factors are the same (same c, same $|v|$); thus by (2.3)

$$T_2 = T_0 K^2, \quad T_3 = T_1 K^2, \quad T_4 = T_2 K^2, \ldots$$

giving two interleaved geometric progressions. Again on account of the symmetry, we have in fact a single progression, with common ratio K:

$$T_n = T_0 K^n. \tag{2.5}$$

2.5 The clock paradox

The times of the pulse echoes as observed by A and B form a geometric progression; in particular,

$$T_n = \sqrt{(T_{n-1} T_{n+1})}.$$

On the other hand, the times *estimated* by A (or B, depending on n) by using the radar formula (2.1) are

$$t_n = (T_{n-1} + T_{n+1})/2.$$

By a well-known result on the arithmetic and geometric means of a pair of positive numbers $t_n > T_n$, true for any n.

Let us spell out precisely what this says:

If n is odd, the event of B's clock showing T_n is estimated by A to occur at the later time t_n: as seen by A, B's clock is slow.

If n is even, the event of A's clock showing T_n is estimated by B to occur at the later time t_n: as seen by B, A's clock is slow.

It follows that there is no hope of synchronizing two clocks in relative motion. This is one version of the celebrated **clock paradox**.

It may need to be emphasized that there is nothing inconsistent here, and that everything follows cleanly from initial assumptions having their reasonable foundation in observation. Nevertheless, it was at this very early point in the development that the critics rejected Relativity. The sticking point has been the intuitive insistence on a universal time, flowing evenly at the same rate at all places in the Universe, with which any local clock may be synchronized. Such a philosophical requirement is not needed by the experimentalist; for him, time is a local matter, it is what his own laboratory clock does, and the performance of other clocks elsewhere and elsewhen is not really relevant.

By how much does a clock go slow? Eliminating T_{n-1} and T_{n+1} from

$$T_n = K T_{n-1}, \quad T_{n+1} = K T_n, \quad t_n = (T_{n-1} + T_{n+1})/2$$

gives

$$t_n = \gamma T_n \tag{2.6}$$

where $\gamma = (K + K^{-1})/2 = (1 - v^2/c^2)^{-1/2}$. This is the relation we are looking for between clock time and coordinate time. This γ has been seen already in connection with the Michelson–Morley experiment and the Fitzgerald–Lorentz contraction, and is the exact version of the approximate result derived through the Principle of Equivalence in section 1.10.

2.6 'Addition' of velocities

Let us turn to three inertial observers A, B and C, moving on the same straight line. B moves away from A with uniform relative velocity u, while C moves away from B with uniform relative velocity v. Since C and A are both inertial, we expect that C moves away from A with some uniform relative velocity w. We know u and v; what is w?

In Fig. 2.4, we assume that A, B and C coincide at $t = 0$; this is not really a restriction. Each observer sees a certain radar pulse at observed times T_A, T_B and T_C respectively; the K-calculus gives immediately by (2.5)

$$T_B = T_A K_u, \quad T_C = T_B K_v, \quad T_C = T_A K_w.$$

Hence

$$K_w = K_u K_v;$$

or, squaring and clearing of fractions (see (2.4)),

$$(c + w)(c - u)(c - v) = (c - w)(c + u)(c + v).$$

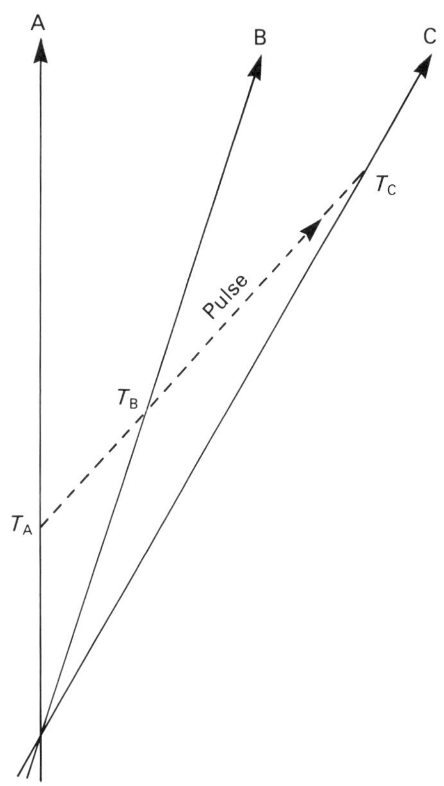

Fig. 2.4

Solving for w:

$$w = \frac{u + v}{1 + uv/c^2}. \tag{2.7}$$

This is very different from the Newtonian formula, $w = u + v$. The reason that we do not notice the difference in everyday life is that everyday velocities are so tiny compared to the velocity of light that the denominator is indistinguishable from unity. It is another matter for high-energy particle physicists, for example, who fling their particles around at much greater velocities. After our experience with the clock paradox, we need not be surprised at this formula; its complicated form reflects the fact that velocities involve time dimensionally, and different observers will be expected to use different relative time-scales.

Equations (2.4), (2.6) and (2.7) suggest that there may be a qualitative difference

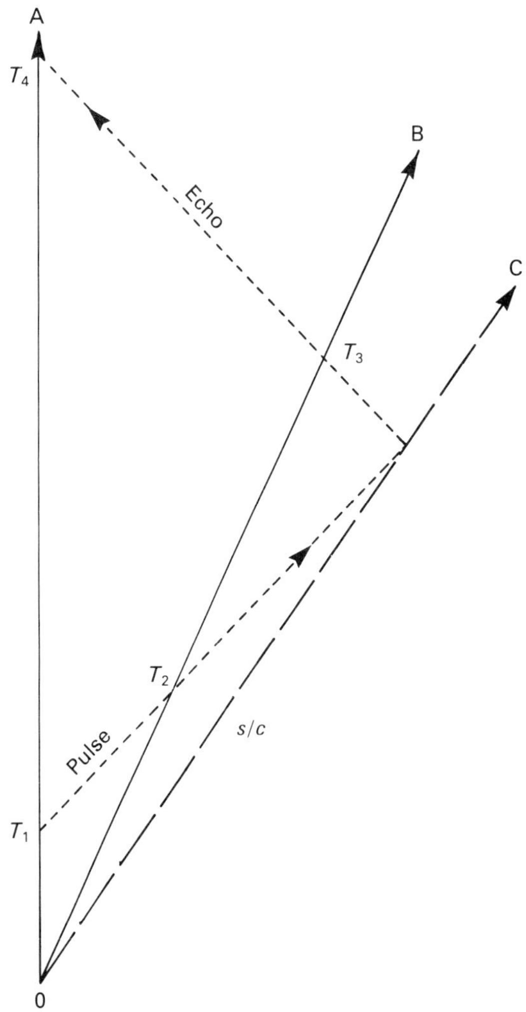

Fig. 2.5

between velocities less than c and those greater than c. In fact, Special Relativity forbids velocities greater than c, while a velocity of c is possible only for certain entities, such as radar pulses. The velocity of any structure with a nonzero mass is necessarily less than c. This will become evident later, when acceleration comes to be discussed.

2.7 Interval

Let us now ask how our observers regard their common environment. Suppose that two inertial observers A and B observe the same event P (Fig. 2.5). We may imagine that A is the one to originate a radar pulse at (his) time T_1. Subsequently, this pulse passes B (at her time T_2), is reflected at P, is seen again by B (T_3), and finally arrives back at A (T_4).

Observer A uses the radar formulas (2.1) to compute his coordinates for the event, and B does the same, obtaining

$$\left.\begin{array}{ll} x_A = c(T_4 - T_1)/2 & x_B = c(T_3 - T_2)/2 \\ t_A = (T_4 + T_1)/2 & t_B = (T_3 + T_2)/2. \end{array}\right\} \tag{2.8}$$

It is important to recognize that there is no reason to suppose that A's assignment is the same as B's.

Additionally we have, by the K-calculus,

$$T_2 = KT_1 \quad \text{and} \quad T_4 = KT_3. \tag{2.9}$$

Equations (2.8) and (2.9) provide six relations, which we may use to eliminate the five variables T_1, T_2, T_3, T_4, and K. Doing this leads to

$$c^2{t_A}^2 - {x_A}^2 = c^2{t_B}^2 - {x_B}^2.$$

This is a remarkable relation. The left side has no reference to B, and the right side none to A. There is no mention of the relative velocity of the observers. We must conclude that the expression

$$c^2t^2 - x^2 \equiv s^2$$

has the same value for every inertial observer whose worldline passes through the event O. It is therefore crucially important, since it is something that all such observers agree on; s happens to be the **interval** between the events O and P.

The interval s may be real, zero, or pure imaginary, depending on the sign of the left side. When it is real, it is possible to have an inertial observer C whose worldline passes through both O and P; in this case, $x_C = 0$, and $s = ct_C$, and thus s/c is the lapse of time between O and P as measured directly by C. This gives a simple observational meaning to the interval. In fact,

interval along any worldline is measured by the clock whose worldline it is.

This interpretation of interval as **proper time** is fundamental. In the sequel, worldlines will regularly be described by their **parametric representation** in which they are to be thought of as being traced out as proper time τ passes; the four coordinates of a typical event on the worldline will be provided by four appropriate functions of τ.

Much more generally, we may consider the interval between any two events P_1 and P_2 in the full (1 + 3)D Minkowski spacetime. Suppose that a chosen inertial observer computes a time and a position for each of the two events: (t_1, x_1, y_1, z_1) and (t_2, x_2, y_2, z_2). Then it may be shown that the expression

$$s^2 = c^2(t_1 - t_2)^2 - (x_1 - x_2)^2 - (y_1 - y_2)^2 - (z_1 - z_2)^2 \tag{2.10}$$

takes the same value, whoever the inertial observer is who computes it, and therefore relates to the two events alone: *s is the interval between the events.* (We shall take this important fact without proof.)

2.8 Spacetime

The Minkowski diagram was introduced originally as nothing more than a plot of events in space and time: being dimensionally different, the space and time were not expected to mix. However, in (2.10), they do mix, and the velocity c plays the part of a conversion factor relating the units of length and of time (1 nanosecond $\equiv$ 0.3 metre, or thereabouts).

The consequences are far-reaching. The diagram is no longer a mere plot; it is a 4D structure which has acquired a **metric**, being bonded together by the network of intervals between pairs of events. For the moment, the spacetime of Special Relativity is *flat*, but the new structure allows for the possibility of the introduction of a *curvature* which is at the root of the Einstein theory of gravity.

It is usual to write the metric in a differential form, with reference to a pair of *neighbouring* events whose coordinates differ by the small amounts dt, dx, dy, dz. The interval ds between such a pair is given by

$$\mathrm{d}s^2 = c^2\,\mathrm{d}t^2 - \mathrm{d}x^2 - \mathrm{d}y^2 - \mathrm{d}z^2. \tag{2.11}$$

2.9 The lightcone

Pick any event P. The **lightcone** (or **null cone**) at P is the 3D collection of all the events Q for which the interval $s(\mathrm{P},\mathrm{Q})$ is zero; the equation of this hypersurface is (when P is taken to be at the origin of coordinates)

$$c^2t^2 - x^2 - y^2 - z^2 = 0. \tag{2.12}$$

(Fig. 2.6 shows the section of the full surface obtained by setting $z = 0$, and is enough to make the main features evident.) The lightcone divides the rest of spacetime into three regions:

(i) The **future** of P—all those events Q for which $s(\mathrm{P},\mathrm{Q})$ is real, and $t(\mathrm{Q}) > 0$;
(ii) The **past** of P—events Q for which $s(\mathrm{P},\mathrm{Q})$ is real, and $t(\mathrm{Q}) < 0$;
(iii) The **elsewhere** of P—events Q for which $s(\mathrm{P},\mathrm{Q})$ is imaginary.

The cone itself has a future and a past part.

The structure of the lightcone has implications for **causality**. Since it appears that the velocity c is an upper limit to the speed at which influences can propagate, no worldline through P can ever wander into the elsewhere of P. Consequently, phenomena at P cannot influence phenomena at Q unless Q *lies in the future of* P

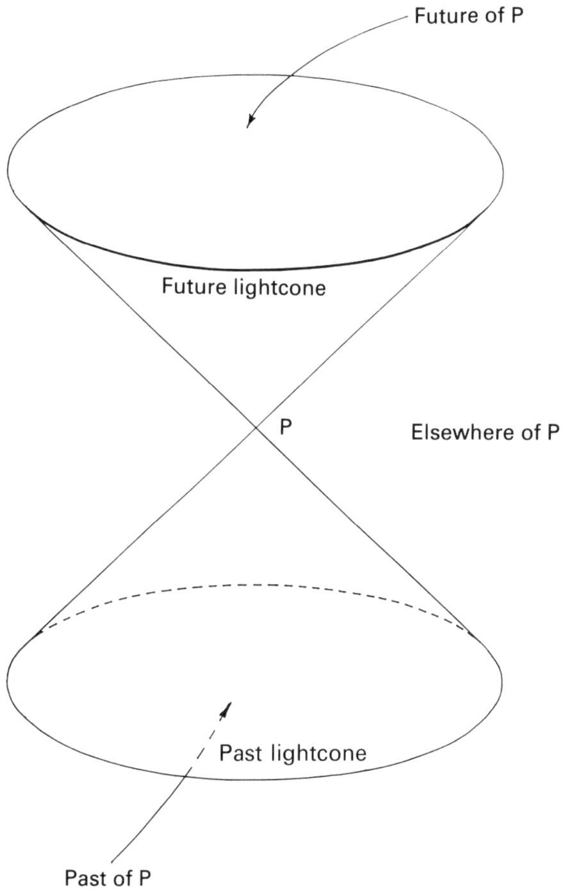

Fig. 2.6

(or perhaps on the future part of the lightcone at P). Similarly, if Q is to influence P in any way, Q must lie in P's past. (There have been discussions as to whether this may be altogether true when quantum mechanics is taken into account, but such matters are far beyond the scope of this book.)

2.10 Signature

Envisage a generalized nD space in which the 'squared-length' of a vector $\mathbf{x} = (x_1, x_2, \ldots, x_n)$ is given by the formula

$$|x|^2 = +x_1{}^2 + x_2{}^2 + \cdots$$
$$\cdots - x_{n-1}{}^2 - x_n{}^2, \tag{2.13}$$

in which there are n_1 positive signs and n_2 negative signs. If the squared-length is zero, then we say that the vector is **null**. The **scalar product** of two vectors $\mathbf{x}$ and $\mathbf{y}$

is defined by

$$\mathbf{x}\cdot\mathbf{y} = +x_1y_1 + x_2y_2 + \cdots$$
$$\cdots - x_{n-1}y_{n-1} - x_ny_n. \tag{2.14}$$

If this scalar product is zero, then we say that the vectors are **mutually orthogonal.**

Now in the collection of n vectors

$$\begin{gathered}(1,0,0,\ldots,0)\\(0,1,0,\ldots,0)\\(0,0,1,\ldots,0)\\\vdots\\(0,0,0,\ldots,1)\end{gathered}$$

every pair of vectors is evidently mutually orthogonal, and there are exactly n_1 vectors with positive squared-length, and n_2 vectors with negative squared-length. This leads us to an important property of the space itself: choose *any set whatever* of n non-null vectors, mutually orthogonal in pairs; then among them there will be precisely n_1 (and n_2) vectors of positive (and negative) squared-length. (We shall take this without proof.)

The pair of numbers n_1 and n_2 is the **signature** of the space; it is an important feature of the space itself; we shall sometimes emphasize the signature by referring to nD space as $(n_1 + n_2)$D space. (Signature is defined by other authors as the difference $n_1 - n_2$.) Some examples are

the plane	$(2 + 0)$D
ordinary 3D space	$(3 + 0)$D
Minkowski spacetime with t and x	$(1 + 1)$D
full Minkowski spacetime	$(1 + 3)$D.

Spaces with different signatures may have very different properties as regards the existence and shape of null surfaces, such as the lightcone.

It is a fundamental feature of Relativity to regard the collection of events which make up the history of the Universe as the 'points' in a $(1 + 3)$D spacetime; this will be just as true when we come to introduce 'curvature' into the theory. There is a mental pitfall in the use of words such as *point* and *curvature*; it is tempting to import ideas and pictures from our familiar 3D space, and to begin to think of the history of the Universe as some kind of 4D *object*, given once for all, without any idea of beginning or development or end. (How can the future be *curved*?) The reason we use the names that we do is that we are using the same mathematical techniques which were used originally for easily visualized curved surfaces. If we are not careful to remember that we are using the words by analogy, we shall run into all kinds of philosophical nonsense.

2.11 The Lorentz transformation

If the interval between a pair of events is the same for any inertial observer, there must exist a transformation carrying one observer into another. Return to the $(1 + 1)$D case of (2.8) and (2.9), eliminate T_1, T_2, T_3, T_4, and solve for t_B and x_B to obtain

$$t_B = \gamma(t_A - (v/c^2)x_A)$$
$$x_B = \gamma(-vt_A + x_A), \tag{2.15}$$

where, as usual,

$$\gamma = (1 - v^2/c^2)^{-1/2}.$$

These formulas allow B's version of events to be obtained from A's version. They form the one-space-dimensional **Lorentz transformation**. Geometrically, the transformation may be taken as a change of coordinate axes in the Minkowski diagram, as shown.

A convenient and easily remembered matrix version of the transformation is

$$\begin{bmatrix} ct_B \\ x_B \end{bmatrix} = \gamma \begin{bmatrix} 1 & -v/c \\ -v/c & 1 \end{bmatrix} \begin{bmatrix} ct_A \\ x_A \end{bmatrix}. \tag{2.16}$$

The complete collection of all possible Lorentz transformations form the **Lorentz group**, characterized as the set of all linear transformations of ct, x, y, and z which leave the formula for $s(\mathrm{P}, \mathrm{Q})$ unchanged. The properties of this group are important to experimenters in the field of, for example, elementary particle physics. They are fairly complicated and, since they are not essential for this book, we shall not consider them further.

2.12 An example

Systematic use of the Lorentz transformation is often the best way of dealing with problems where intuition may otherwise lead us astray. The following is typical.

A spaceship B of restlength L passes an observer A, who measures the time T for the ship to pass him. What is the speed v of B with respect to A?

There are two events to be considered:

Event 1—The nose of B passes A;

Event 2—The tail of B passes A.

It is convenient to place the origin of coordinates—for both A and B—at the event 1. Our information so far may be summarized as follows:

	In A's coordinates	In B's coordinates
Event 1	$(0, 0)$	$(0, 0)$
Event 2	$(T, 0)$	$(T^?, -L)$.

(Of course, we are told nothing about the time denoted by $T^?$. It would be wrong to expect it to be the same as T.) These coordinates are related by the Lorentz transformation for the (unknown) velocity v. In particular,

$$\begin{bmatrix} cT^? \\ -L \end{bmatrix} = \gamma \begin{bmatrix} 1 & -v/c \\ -v/c & 1 \end{bmatrix} \begin{bmatrix} cT \\ 0 \end{bmatrix}.$$

The lower line of this relation is $L = \gamma v T$; solving for v gives the result

$$v = L/\sqrt{(T^2 + L^2/c^2)}.$$

When c is relatively very large, this goes over into the everyday relation $v = L/T$. The paradoxes of Special Relativity may be thought of as proceeding from an overenthusiastic insistence that this everyday relation should somehow be universally true. Thus the actual relation $L = \gamma v T$ may be rearranged as $v = (L/\gamma)/T$: the everyday relation for a Fitzgerald-contracted L and an uncorrected time T; this is A's view of things. Alternatively, $v = L/(\gamma T)$: this corresponds to B's view of an uncorrected L and a dilated T.

It would seem best to avoid such hybrid rearrangements.

2.13 Acceleration

We must now turn to consider non-inertial observers, and ask: How is the reading of the accelerometer of a non-inertial observer B related to the course of his worldline? Specifically, B makes a continuous record of his own accelerometer reading $f(\tau)$ as a function of his own proper time τ; how can he use his observations to plot his worldline in the Minkowski diagram of A? (This is a very practical problem which lies at the root of all so-called inertial navigational systems.) We shall do the work for one space dimension only.

An accelerometer measures rate of change of velocity. Thus, during a small interval of proper time $\delta\tau$, B will reckon that his velocity has changed by an amount $\delta u = f(\tau)\delta\tau$. From his different viewpoint, A sees B moving with a nonzero velocity $V(\tau)$; after the interval $\delta\tau$, he will see the new velocity

$$V(\tau + \delta\tau) = (V + \delta u)/(1 + V\delta u/c^2)$$

according to the law (2.7) for summing two velocities. To first order of smallness in δu,

$$V(\tau + \delta\tau) - V(\tau) = \delta u(1 - V^2/c^2);$$

hence

$$\dot{V} \equiv \frac{dV(\tau)}{d\tau} = \left(1 - \frac{V^2}{c^2}\right) f(\tau). \tag{2.17}$$

This is a differential equation for V when the observed acceleration is given as the function $f(\tau)$. (Note that *we shall always use the dot notation—$\dot{V}$—to signify differentiation with respect to a parameter, usually proper time along a worldline.*) A slight rearrangement gives

$$\frac{dV}{1 - V^2/c^2} = f(\tau)\, d\tau,$$

which integrates easily to

$$V(\tau) = c \tanh \phi(\tau), \tag{2.18}$$

where

$$c\phi(\tau) = \int_0^\tau f(\tau)\, d\tau$$

in which $V(\tau = 0)$ is zero; see Problem 12 for the more general case. (This essentially gives the relation between the *direction* of the worldline, as specified by V, as a function of the proper time parameter τ. It is sometimes known as the **intrinsic equation** for the curve.)

To proceed further, we use the timekeeping relation $\dot{t} = \gamma$ in the form

$$\dot{t} \equiv \frac{dt}{d\tau} = \frac{1}{\sqrt{(1 - V^2/c^2)}} = \cosh \phi(\tau),$$

while

$$\dot{x} \equiv \frac{dx}{d\tau} = \frac{V}{\sqrt{(1 - V^2/c^2)}} = c \sinh \phi(\tau).$$

Integration of the last two equations with respect to τ will give the functions $t(\tau)$ and $x(\tau)$. As τ varies, the values $(t(\tau), x(\tau))$ trace out the worldline of B in A's diagram; this is the parametric representation of the worldline mentioned earlier. From the point of view of observer B, the parametric representation is very natural: it tells him exactly what he wants to know—where he is on the 'map' when his clock registers τ. It is something he can calculate for himself from his own accelerometer readings, without reference to the outside world.

2.14 Four-velocity and four-acceleration

The three components of the conventional 3D-velocity that we have used so far are

$$\left(\frac{dx}{dt}, \frac{dy}{dt}, \frac{dz}{dt}\right) = (v_x, v_y, v_z).$$

The corresponding **four-velocity** is defined by

$$\dot{\mathbf{x}} = (c\dot{t}, \dot{x}, \dot{y}, \dot{z}); \tag{2.19}$$

it is a **four-vector** which is in fact the **unit tangent** to the worldline in Minkowski spacetime. The relation between the two is easily seen to be

$$(c\dot{t}, \dot{x}, \dot{y}, \dot{z}) = \left(c, \frac{dx}{dt}, \frac{dy}{dt}, \frac{dz}{dt}\right)\frac{dt}{d\tau} = (c, \mathbf{v})\gamma.$$

The **squared-length** of any four-vector is obtained via the interval formula; for the four-velocity it is

$$\dot{\mathbf{x}} \cdot \dot{\mathbf{x}} = c^2\dot{t}^2 - \dot{x}^2 - \dot{y}^2 - \dot{z}^2 = (c^2 - \mathbf{v}^2)\gamma^2 = c^2.$$

It is in this sense that the four-velocity is a unit vector (the factor of c reflects the most natural choice of units for a velocity).

The **four-acceleration** is the rate of change of the four-velocity:

$$\mathbf{F} = \ddot{\mathbf{x}} = (c\ddot{t}, \ddot{x}, \ddot{y}, \ddot{z}).$$

It is a four-vector for the same reason that the four-velocity is a four-vector. Its squared-length is of interest: for the 1D motion of the last section, for example, differentiating the four-velocity gives

$$\mathbf{F} = c\dot{\phi}(\sinh \phi, \cosh \phi, 0, 0), \tag{2.20}$$

the squared-length of which is $-c^2\dot{\phi}^2$. The negative sign shows that the four-acceleration, unlike the four-velocity, is a spacelike vector: it always points into the 'elsewhere' in the Minkowski spacetime. (Moreover, the four-acceleration is always orthogonal to the four-velocity.) Apart from the sign, the magnitude of the acceleration is $c\dot{\phi}(\tau) = f(\tau)$; this is what the accelerometer indicates.

2.15 Event horizons

The presence of acceleration may produce a surprising effect. Consider for simplicity the case of *uniform* acceleration:

$$f = \text{constant} \quad \text{and} \quad \phi = f\tau/c.$$

The differential equations are now easily solved to give

$$t = t_0 + (c/f)\sinh f\tau/c,$$
$$x = x_0 + (c^2/f)(\cosh f\tau/c - 1),$$

where t_0 and x_0 are constants of integration; an appropriate choice gives the instance shown in Fig. 2.7. The worldline takes the form of a *hyperbola.*

Imagine that the observer has the resources to maintain an acceleration f for ever. Then there will be events that he will never be able to observe. The events in question lie on the future side of the asymptote of the hyperbola; the asymptote itself is a null line, and is the **event horizon** of the accelerated observer. Objects whose worldlines cross the horizon will disappear from the observer's view, and will seem to take for

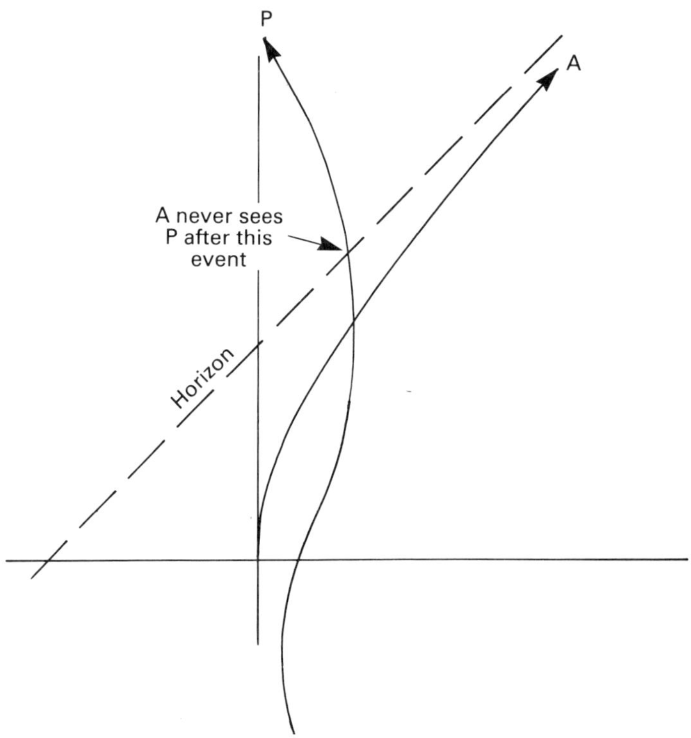

Fig. 2.7

ever to do so. Nevertheless, the objects themselves cross the horizon in a finite proper time, and will still have an infinite lifetime ahead of them.

The event horizon is in no sense a barrier, but rather an artefact of the acceleration of the observer. We shall meet event horizons again when we come to consider 'black holes'.

2.16 Who did it all?

Most people believe that Special Relativity was the work of Einstein alone. This is not entirely so. The time was ripe for the discovery of Relativity in any case. In 1900, for example, Larmor had already suggested that moving clocks must run slow, and by how much. Lorentz published the final version of his transformation in 1903 (Poincaré was the one responsible for attaching the name of Lorentz to it.) Henri Poincaré was perhaps the most far-seeing of the thinkers at the time. For example, in 1899 he asserted his belief that absolute motion is undetectable in principle by any means whatever; he was really recalling the faithful to the original Relativity of Galileo and Newton. In 1904, he had moved to the point of saying that there must be a new dynamics, in which no velocity can exceed that of light. Max Planck did much to determine what that mechanics might be.

So the foundation was well and truly laid when Einstein picked up the torch in 1905. And if he hadn't done it, someone else would have.

General Relativity is another matter. Einstein has acknowledged his debt to Minkowski: the recognition that his diagram was properly a 4D space gave Einstein the clue he needed to develop a full theory of gravity. From this point on, it seems that the work is his own.

Lorentz is an interesting figure. In spite of his early involvement, he was never really happy with Relativity, and throughout his life feft that space and time are completely distinct and that a universal 'true' time exists. Interestingly, recent delicate experiments in quantum theory have raised some philosophical difficulties for Relativity, and it has been commented (by Karl Popper, I believe) that a return to a Lorentz point of view may become necessary. But this is not the place!

2—Notes and problems

1. Two identical clocks A, B, move with constant relative velocity v. They meet momentarily, and are synchronized to zero. Later, at time T_A, A emits a pulse which is received by B at time T_B. Show that

$$v = c\frac{{T_B}^2 - {T_A}^2}{{T_B}^2 + {T_A}^2}.$$

   (Hint: Rearrange $T_B = KT_A$.)

   What is the corresponding formula for a pulse emitted by B and received by A?

2. Show that the arithmetic mean of two unequal positive numbers x, y is greater than their geometric mean. (Use $[\sqrt{x} - \sqrt{y}]^2 > 0$.)

3. Four observers A, B, C, and D move on the same straight line with the uniform relative velocities: u (B with respect to A), v (C to B), and w (D to C). Show that

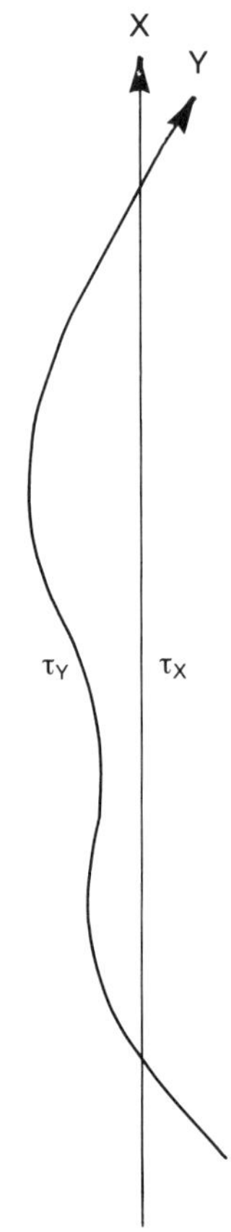

Fig. 2.8

the velocity of D with respect to A is

$$\frac{u + v + w + uvw/c^2}{1 + (uv + uw + vw)/c^2}.$$

(Consider the product of three K-factors, or else apply the formula for the composition of two velocities twice.)

4. **The twin paradox** It is traditional to present the clock paradox in a rather fanciful way as allows. X and Y are twins. X stays at home (remaining inertial at all times), while Y goes on an extended space flight. When Y returns, he looks younger, feels younger, and indeed *is* younger—if the evidence of his clock is to be believed—than his brother X.

   Plot the course of events in the Minkowski diagram of X. (Fig. 2.8). The worldline of X is the t-axis, while that of Y is some curve—parametrically given as $t(\tau)$, $x(\tau)$, $y(\tau)$, $z(\tau)$—which starts and ends on the t-axis, at t_0 and t_1, say. The lapse of proper time for X is simply $\tau_X = t_1 - t_0$, while that for Y is the integrated interval

$$\tau_Y = \int \sqrt{(\dot{t}^2 - \dot{x}^2 - \dot{y}^2 - \dot{z}^2)}\, d\tau.$$

Show that $\tau_Y < \tau_X$, and thus that Y ends up younger than X. (Hint: it is evident that

$$\dot{t} \geqslant \sqrt{((\dot{t}^2 - \dot{x}^2 - \dot{y}^2 - \dot{z}^2).)}$$

It is sometimes suggested that this result is due to some mysterious effect of Y's acceleration on Y's clock. True, some clocks (pendulum clocks, for example) are catastrophically susceptible to accelerations, but this is emphatically not to the point. The formula for the integrated interval involves velocities only, not accelerations.

It has occasionally been argued that Relativity supposes a perfect symmetry between any pair of observers like X and Y, and therefore that when they meet neither can be slow with respect to the other. But Relativity says no such thing, and to suggest it implies a fundamental misunderstanding of the theory. X and Y are *not* mutually symmetric: X is inertial; Y is not.

5. An astronaut wants to go to a star 10 light years away. The rocket accelerates rapidly, and then moves with uniform velocity. What velocity is needed if the astronaut is to arrive after a proper time of one year? (Note: $\gamma = 0.1$.)

   Is the journey feasible?

6. The event A $(a_0, \mathbf{a})$ lies on the worldline of an inertial observer who has previously visited the origin O. Show that the observer's 'plane surface' (actually a 3D space) of simultaneity through A has equation

$$c^2 a_0 t - \mathbf{a}\cdot\mathbf{r} = c^2 {a_0}^2 - \mathbf{a}^2$$

   where $\mathbf{r} = (x, y, z)$. (Hint: If B is any event on this surface, AB is to be orthogonal to OA.)

7. **A wavefront that is spherical for one observer is spherical for all** A pulse of light is emitted at the origin; the worldlines of the photons therefore lie on the null cone

$$c^2 t^2 - \mathbf{r}^2 = 0.$$

   A typical wavefront (for our observer) is the intersection of this null cone with the plane of simultaneity through A. Show that for any point $(t, \mathbf{r})$ on this intersection,

$$c^2(t - a_0)^2 - (\mathbf{r} - \mathbf{a})^2 = -(c^2 {a_0}^2 - \mathbf{a}^2).$$

   Thus, all points on the wavefront are equidistant from A.

   There is no paradox, even though the result is true for arbitrary choice of the observer, since the set of events which constitutes a simultaneous wavefront for one observer *cannot* be the same as a comparable set for another, relatively moving, observer.

8. The composition of non-collinear velocities begins to involve the full panoply of the Lorentz group, and this forms no part of the programme of this book. However, here is a fairly straightforward subsidiary result for general velocities.

Inertial observer A watches two other inertial observers B and C, moving with the relative (non-collinear) velocities **U** and **V**. Moreover, the worldlines of the three observers are concurrent, at the origin of A's diagram, say. What is the mutual relative speed w of B and C?

A sketch of the solution follows. The worldline of B in A's Minkowski plot is, parametrically,

$$t_B = \gamma_U \tau, \quad \mathbf{r}_B = \gamma_U \mathbf{U}\tau.$$

The worldline of C is given similarly. B sends out a pulse at time τ_B, and C receives the pulse at time τ_C. The fact that the worldline of the pulse is null yields

$$c^2(\gamma_V \tau_C - \gamma_U \tau_B)^2 - (\gamma_V \mathbf{V}\tau_C - \gamma_V \mathbf{U}\tau_B)^2 = 0.$$

Also, $\tau_C = K\tau_B$, where K is the mutual K-factor of B and C. Eliminate τ_B and τ_C to obtain

$$K^2 - 2\gamma_U\gamma_V(1 - \mathbf{U}\cdot\mathbf{V}/c^2)K + 1 = 0.$$

Use $K^2 = (c + w)/(c - w)$ to show finally that

$$w^2 = \frac{(\mathbf{U} - \mathbf{V})^2 - (\mathbf{U} \wedge \mathbf{V})^2/c^2}{(1 - \mathbf{U}\cdot\mathbf{V}/c^2)^2}.$$

When the worldlines of the observers are not either coplanar or concurrent, the analysis is much more complicated.

9. The half-life of a charged pion is 1.77×10^{-8} s. A pion beam leaves an accelerator with a speed $0.99c$. Over what distance D will the intensity of the beam drop by one-half?

As a first step, notice that we are concerned with the events:

Event 1—A pion leaves the accelerator;

Event 2—The pion reaches the half-life point.

It is useful to draw up the information relating to observers A (at rest in the laboratory) and M (moving with the meson). Thus,

	Observer A	Observer M
Event 1	$(0, 0)$	$(0, 0)$
Event 2	$(?, D)$	$(1.77 \times 10^{-8}\,\mathrm{s}, 0)$.

Now apply the Lorentz transformation with $v = 0.99c$ to obtain $D = 37.1$ m. (Note that this is more than seven times further than the nonrelativistic formula gives, a difference which is experimentally very well substantiated.)

10. A spaceship A of restlength L overtakes a second (longer) spaceship B at a relative speed v and on a parallel course. As they pass, a crew member of A fires simultaneously (in his reckoning) a very short burst from each of two lasers mounted transversely at the nose and tail of A. Calculate the separation of the resulting burnmarks on B.

This time the relevant tabulation is

		A	B
1	Front laser firing	$(0, L)$	$(?, S)$
2	Rear laser firing	$(0, 0)$	$(0, 0)$

which will lead to $S = \gamma L$.

It is important to understand that when the spaceships return to rest at base, the separation of the burnmarks on B will be found to be genuinely greater than the separation of the lasers on A. A will explain that the Fitzgerald shrinkage of B is responsible. B claims that, even though A appears to him to be the one who has shrunk, the firings are not simultaneous, the front firing lagging so much as to *over*compensate for A's deficiency in length (to see this, calculate the ? in the above tabulation). Of course, the captain of B, being a good relativist, will not blame this on the incompetence of the crew of A.

11. Return to Problem 1. Here there are three events to be considered:

		A's coordinates	B's coordinates
1	A and B separate	$(0, 0)$	$(0, 0)$
2	A transmits pulse	$(T_A, 0)$	?
3	B receives pulse	$(cT_A/(c-v), cvT_A/(c-v))$	$(T_B, 0)$.

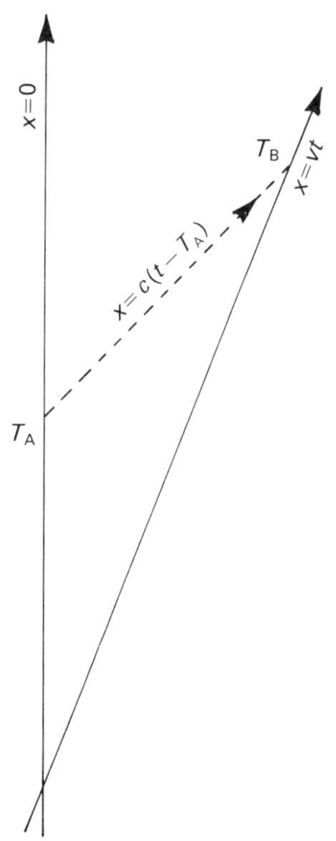

Fig. 2.9

The least obvious entries are found by finding the point of intersection of two worldlines whose equations are known; see Fig. 2.9.

Show that applying the Lorentz transformation to event 3 leads to $T_B = KT_A$. (This does nothing more than to confirm the work of the main text. After all, the Lorentz transformation was originally founded on the K-factor relation.)

12. Show that the general solution of

$$\frac{dV}{1 - V^2/c^2} = f(\tau)\,d\tau$$

is

$$\operatorname{argtanh}\frac{V(\tau)}{c} - \operatorname{argtanh}\frac{V_0}{c} = c^{-1}\int_0^{\tau} f(\tau)\,d\tau \equiv \phi(\tau),$$

in which V_0 is the constant of integration. Show that this result may be rearranged as

$$\frac{V(\tau) - V_0}{1 - V(\tau)V_0/c^2} = c\tanh\phi(\tau).$$

Relate this final version to the 'addition formula' for velocities.

13. Show that, no matter how large the acceleration, the velocity of a material object can never reach c. (Hint: $-1 < \tanh x < 1$ for all x.)

14. A rocket leaves base at time $t = 0$ and travels for ever in a straight line with uniform acceleration f with respect to its own reference frame. Show that no signal sent from base later than time $t = c/f$ can ever overtake the rocket.

15. An astronaut wants to go to a star 10 light years away, with *uniform acceleration*. What acceleration is needed if the astronaut is to arrive after a proper time of one year? (Note: One solution of $10z = \cosh z - 1$ is $z = 4.53$.)

Is the journey feasible?

16. Show that the four-velocity and the four-acceleration are always mutually orthogonal. [Apply the scalar product (2.14) to the expressions (2.19) and (2.20).]

3

Inertial observers in a curved spacetime

The fundamental feature of Einstein's approach to gravity is that the flat $(1+3)$D spacetime of Minkowski is to be replaced by a curved spacetime, and that the curvature is to be responsible for gravitational effects. Events are still to be represented by 'points' in the spacetime, but relationships between different events and between different inertial observers will be altered on account of the curvature. This chapter will be devoted to developing techniques which will enable us to handle curved spaces properly. In particular, the methods will be *intrinsic*: spacetime needs to be described without reference to any external scaffolding—such as embedding in a flat space of higher dimension. Such scaffolding is unnecessary and may be misleading.

The mathematics is known as **differential geometry**, and has developed prolifically in recent years. It is tempting to explore some of the byways. However, this is a book about gravity physics, and the full panoply of differential geometry will not be needed. Readers who are familiar with the more mathematical aspects will become aware of how much is left out. Only the bare minimum is presented here, and if more is wanted it will be necessary to look elsewhere.

3.1 Coordinates, vectors, and the metric tensor in two dimensions

To talk at all about the points in a space we must be able to name them in some way. It is usual to set up a coordinate system on the space—or in part of the space; it is then straightforward to specify any point by giving the coordinates of that point. For example, polar coordinates r, θ will provide a labelling of the points in a plane. Of course, the choice of a coordinate system is never unique: for a plane, Cartesian coordinates x, y will do just as well. The proper choice of appropriate coordinates is an important matter—some easy problems may become very difficult in unsuitable coordinates. (There are some technical requirements which we shall always take for granted, mainly to do with the 'smoothness'—differentiability—of the system of coordinates we choose.)

The 2D case of a curved surface is easy to picture, and will be dealt with first. On a plane, a finite displacement is a vector with a length and a definite direction. On a curved surface, the idea of a definite direction throughout an extended area becomes imprecise, and finite displacements can no longer be regarded as vectors. However, infinitesimal displacements are still vectors, for the following reason. Envisage a small region excised from a smoothly curved surface in the vicinity of a point P. The geometry of such a region is to be taken as almost the same as the geometry of a similar small region excised from an appropriate plane (the **tangent plane** at P)—if we make the regions small enough, the curvature is hardly noticeable. In fact, to a good approximation, much of flat geometry applies 'in the small', and this will often be a useful guide. In particular, infinitesimal displacements behave as vectors to first order of smallness.

The kingpin of Euclidean geometry is Pythagoras' theorem: the squared-length of a small displacement is an expression which is homogeneous and quadratic in the coordinate changes. For Cartesian coordinates on a plane, this expression is just the well-known sum of squares; the more general statement covers all kinds of coordinate systems, even on curved surfaces. Additionally, it covers spaces with a different signature, such as Minkowski spacetime of events, where 'squared-length' is to be replaced by *squared-interval*, which may be zero or even negative.

Suppose that a curved surface is supplied with a curvilinear coordinate system u, v (Fig. 3.1). Any small vector displacement $d\mathbf{r}$ at the point P may be expressed as a linear combination of the corresponding small coordinate changes du and dv,

$$d\mathbf{r} = \mathbf{a}\,du + \mathbf{b}\,dv$$

by the usual Euclidean parallelogram rule, valid 'in the small'. (The vectors $\mathbf{a}$, $\mathbf{b}$ are clearly fixed by the properties of the coordinate grid at P. They are *finite* vectors, and really reside in the tangent plane at P.) The scalar product of two such vectors,

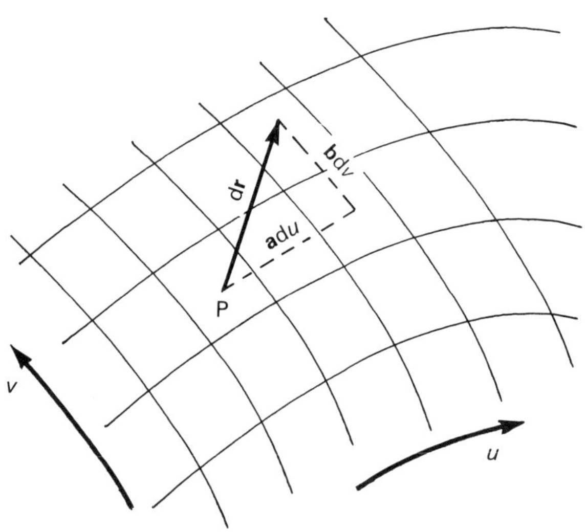

Fig. 3.1

$\mathbf{dr}_1$, $\mathbf{dr}_2$ (both of them will have to be at P), is given in the usual way,

$$\begin{aligned}\mathbf{dr}_1\cdot\mathbf{dr}_2 &= (\mathbf{a}\,du_1+\mathbf{b}\,dv_1)\cdot(\mathbf{a}\,du_2+\mathbf{b}\,dv_2)\\ &= [du_1\;dv_1]\begin{bmatrix} a^2 & \mathbf{a}\cdot\mathbf{b}\\ \mathbf{a}\cdot\mathbf{b} & b^2\end{bmatrix}\begin{bmatrix} du_2\\ dv_2\end{bmatrix}\\ &= [du_1\;dv_1]\mathbf{g}\begin{bmatrix} du_2\\ dv_2\end{bmatrix}\end{aligned}$$

in a convenient matrix form. The elements of the symmetric matrix $\mathbf{g}$—labelled in the natural way—

$$g_{uu}=a^2,\quad g_{uv}=g_{vu}=\mathbf{a}\cdot\mathbf{b},\quad g_{vv}=b^2$$

are fixed by the properties of the coordinate grid at P, and in general are expected to vary from place to place. The g_{ij}, taken together, are the components of the **metric tensor** at P.

It is inevitable that in curvilinear coordinates the simplicity of the Cartesian version of Pythagoras' theorem disappears. On this account, it becomes convenient to specify a vector by its **contravariant** components; on a small displacement $\mathbf{dr}$ these are the coordinate changes du, dv. In general, any vector $\mathbf{V}$ at the point P (velocity, force, etc.) is related to its contravariant components $(V_\mathrm{a}, V_\mathrm{b})$ by the linear relation

$$\mathbf{V}=\mathbf{a}V_\mathrm{a}+\mathbf{b}V_\mathrm{b}.$$

(The contravariant components of $\mathbf{a}$ itself are clearly $(1,0)$. Of course, this does not imply that $\mathbf{a}$ is a unit vector.)

For example, the surface of a sphere (radius R) is supplied with coordinates colatitude θ and longitude ϕ. At the point $\mathrm{P}(\theta,\phi)$, a small change $d\theta$ in colatitude produces a shift $R d\theta$ on the surface, while a small change $d\phi$ in longitude produces a shift $R\sin\theta\,d\phi$. These shifts are orthogonal in this case; by Pythagoras, therefore, the shift ds which results from the infinitesimal displacement with contravariant components $(d\theta, d\phi)$ is given by

$$ds^2=(R\,d\theta)^2+(R\sin\theta\,d\phi)^2=R^2\,d\theta^2+R^2\sin^2\theta\,d\phi^2,$$

from which we may read off the elements of $\mathbf{g}$ at P,

$$g_{\theta\theta}=R^2,\quad g_{\theta\phi}=g_{\phi\theta}=0,\quad g_{\phi\phi}=R^2\sin^2\theta$$

(using an obvious labelling convention).

3.2 The general formalism for *n*D space

Label the points of a (possibly) curved *n*D space with some chosen system of coordinates $x_1, x_2, \ldots, x_n$. The **contravariant components** of a small displacement $\mathbf{dr}$ at the point P are the coordinate changes dx^μ $(\mu=1,\ldots,n)$ which result from the displacement. The length ds of this displacement is given by a quadratic expression (the **line element** or the **metric**)

$$ds^2=\sum_{\mu=1}^{n}\sum_{\nu=1}^{n}g_{\mu\nu}(\mathrm{P})\,dx^\mu\,dx^\nu,$$

in which the n^2 coefficients $g_{\mu\nu}(\mathrm{P})$, taken together, constitute the **metric tensor g** at P. It is to be expected that these coefficients vary from place to place because the properties of the coordinates, or of the space, or of both, vary from place to place. Always, **g** is to be taken to be symmetric: $g_{\mu\nu} = g_{\nu\mu}$; the metric tensor at P thus has $n(n+1)/2$ independent components.

It is a remarkable fact that if we are given the metric throughout a space then in principle we can work out almost everything else. ('Almost': the line element is essentially a 'local' feature, and there may exist 'global' differences between two spaces with the same line element. For instance, an infinite plane and an infinitely long circular cylinder are both *intrinsically* flat surfaces, since each may be constructed from cardboard without stretching or tearing; they are globally different, however.) On the other hand, the form of the line element inevitably depends both on the shape of the space itself and on the coordinate system used to label the space; we shall have to learn to disentangle the two kinds of information.

An infinitesimal displacement is not the only kind of vector. Many geometrical entities relate to a definite point P, and have both magnitude and direction there: such an entity **V** is also a vector, and is specified by its contravariant components V^μ ($\mu = 1,\ldots,n$). The squared-length of **V** is then given by

$$|\mathbf{V}|^2 = \sum_\mu \sum_\nu g_{\mu\nu} V^\mu V^\nu.$$

Moreover, the **scalar product** of two vectors, **U** and **V** (both of course at the same point P), is given by

$$\mathbf{U}\cdot\mathbf{V} = \sum_\mu \sum_\nu g_{\mu\nu} U^\mu V^\nu.$$

3.3 The summation convention

We almost universally write

$$\mathbf{U}\cdot\mathbf{V} = g_{\mu\nu} U^\mu V^\nu, \tag{3.1}$$

adopting the **summation convention** which says that *any affix which appears in any term both as a superfix and as a suffix is to be summed over*. (In equation (3.1) this convention is to be applied both to μ and to ν.)

It is convenient to have an alternative equivalent representation of a vector. The **covariant components** of a vector **V** at the point P are defined by

$$V_\mu = \sum_{\nu=1}^{n} g_{\mu\nu}(\mathrm{P}) V^\nu \quad \text{for } \mu = 1,\ldots,n.$$

(Do not be deceived by the tiny notational difference between V^μ and V_μ. They are the same only for Cartesian coordinates in flat space.) The last equation is, by convention, abbreviated to

$$V_\mu = g_{\mu\nu} V^\nu, \tag{3.2}$$

where we are to understand that this represents the n equations obtained by allowing μ to range from 1 to n in turn. So we here have n equations, each with one term on the left and n terms on the right. In general, the summation convention says

(i) repeated affixes in any term are to be summed over;
(ii) non-repeated affixes are to be assigned all possible values in turn.

The change from V^μ to V_μ given in (3.2) is called **lowering a superscript**. The reverse process, **raising a subscript**, is

$$V^\mu = g^{\mu\nu} V_\nu \quad \text{(summed, of course),} \tag{3.3}$$

in which a *new* set of n^2 coefficients has been introduced. Of course, to be able to get back to where we started, we must have

$$g^{\mu\alpha} g_{\alpha\nu} = \delta^\mu_\nu \equiv \begin{cases} 1 & \text{if } \mu = \nu \\ 0 & \text{if } \mu \neq \nu \end{cases} \tag{3.4}$$

(summed over α, and true for each μ, ν, by the convention). Here there are n^2 linear equations for the n^2 unknowns $g^{\mu\nu}$; they show that the matrix $\mathbf{g}^{-1}$ formed from the elements $g^{\mu\nu}$ in the obvious way is exactly the inverse of the matrix $\mathbf{g}$ formed from the $g_{\mu\nu}$:

$$\mathbf{g}^{-1}\mathbf{g} = \mathbf{I} \quad \text{(the unit matrix).} \tag{3.5}$$

What happens if we raise both suffixes of $g_{\mu\nu}$? Clearly what we want is

$$g^{\mu\alpha} g^{\nu\beta} g_{\alpha\beta},$$

which is $g^{\mu\nu}$, in fact. Thus $g^{\mu\nu}$ is not only the 'inverse' of $g_{\mu\nu}$; it is also the alternative representation of the metric tensor with both suffixes raised. In fact, there are three representations of the metric, the covariant, contravariant and mixed versions respectively. To these there correspond different expressions for the scalar product

$$\mathbf{U}\cdot\mathbf{V} = g_{\mu\nu} U^\mu V^\nu \quad \text{or} \quad g^{\mu\nu} U_\mu V_\nu \quad \text{or} \quad \delta^\mu_\nu U_\mu V^\nu \equiv U_\mu V^\nu.$$

At a chosen point P, scalar products (and therefore orthogonality and squared-length) are now defined by the metric tensor at P, which must therefore determine the *signature* of the space at P. In fact, if the $n \times n$ matrix $\mathbf{g}$ has n_1 positive and n_2 negative eigenvalues, then the space at P is of signature $(n_1 + n_2)$D (see Problem 9). On account of the subject matter, our concern is naturally with spacetimes of signature $(1+3)$D. However, other signatures will appear, either in examples of the methods, or in subspaces of the full spacetime which will need to be discussed on occasion.

The signature of spacetime cannot sensibly change value from region to region. At any point where such a change might take place the mathematical structures collapse; in any case, all our desirable physical requirements will be violated.

3.4 Changing the coordinates

Suppose that two overlapping coordinate systems are in use; in the region of overlap the systems will be related, and we may wish to change from one system to the other.

Consider, for example, the plane, supplied with Cartesian coordinates and with polar coordinates. In Cartesian coordinates the metric takes the simple Pythagoras form

$$\mathrm{d}s^2 = \mathrm{d}x^2 + \mathrm{d}y^2;$$

what is the metric in polar coordinates?

We shall transform from the 'old' coordinates x, y to the 'new' coordinates r, θ in the following way. First, it is usually more convenient to write the old coordinates in terms of the new, when possible; in this case,

$$x = r\cos\theta, \quad y = r\sin\theta.$$

These relations are to be differentiated in the form

$$\begin{aligned} dx &= dr\cos\theta - r\,d\theta\sin\theta, \\ dy &= dr\sin\theta + r\,d\theta\cos\theta. \end{aligned}$$

Substitution in the old form of the metric now gives

$$\begin{aligned} ds^2 &= dx^2 + dy^2 \\ &= [dr\cos\theta - r\,d\theta\sin\theta]^2 + [dr\sin\theta + r\,d\theta\cos\theta]^2 \\ &= dr^2 + r^2\,d\theta^2 \end{aligned}$$

on multiplying out, and noting that the cross terms involving $dr\,d\theta$ cancel.

It should be clear how we could have obtained this result directly by considering the nature of polar coordinates in the plane. Things are not always as obvious as this, however, and such techniques for coordinate transformation will be needed later.

The relation connecting (dx, dy) and $(dr, d\theta)$ may be written

$$\begin{bmatrix} dx \\ dy \end{bmatrix} = \begin{bmatrix} \cos\theta & -r\sin\theta \\ \sin\theta & r\cos\theta \end{bmatrix} \begin{bmatrix} dr \\ d\theta \end{bmatrix}$$

and this in fact specifies how the contravariant components of *any* vector $\mathbf{A}$ are to change:

$$\begin{bmatrix} A^x \\ A^y \end{bmatrix} = \begin{bmatrix} \cos\theta & -r\sin\theta \\ \sin\theta & r\cos\theta \end{bmatrix} \begin{bmatrix} A^r \\ A^\theta \end{bmatrix}.$$

All entities which represent genuine geometrical features (such as vectors) in the plane itself transform according to this—or some related—scheme. Coordinate-dependent features do not. In general, a **tensor** is a geometrical entity $\mathbf{T}$ whose manifestation in a particular coordinate system is as a collection of **elements** $T\ldots$, indexed by **superfixes** (the contravariant affixes) and **suffixes** (the covariant affixes). Changing the coordinates entails a homogeneous linear transformation of the entire set of elements according to a precise transformation law; the scheme just given is one of the simpler examples.

This is an important matter. All the essential relationships of General Relativity are *tensor* relationships, since these have the persistence against coordinate changes that is necessary for a truly *physical* relationship.

3.5 Geodesics

In Newtonian mechanics the trajectory of a free particle is a *straight* line, traversed with constant velocity. In Special Relativity, an inertial observer with a four-velocity $\gamma(c, \mathbf{v})$ has a worldline—using proper time τ as parameter—

$$\begin{aligned} t &= \gamma\tau, \\ \mathbf{r} &= \gamma\mathbf{v}\tau + \text{constant}, \end{aligned}$$

that is, a straight line in Minkowski spacetime. What should an inertial observer do in a spacetime which may be curved?

The fundamental replacement for Newton's law is

> **The worldline of an inertial observer is to be a *geodesic*, that is, a *straightest possible worldline.* The clock of such an observer registers *proper time*, namely the interval measured along the worldline.**

We are not quite ready to discuss 'straightness', and we shall approach the idea of a geodesic by a different route: *a* **geodesic** *is a curve of stationary length.* The 'straightest' line gives the 'shortest' distance between two points. (We have to be careful: sometimes it can be the longest. The shortest route from London to New York is along the great circle joining them. The longest equally straight route is round the complementary segment of the same great circle.)

3.6 The Lagrangian recipe for a geodesic

We represent a curve by the movement of a point $\mathbf{x}(\sigma)$ as a parameter σ varies. The infinitesimal displacement

$$\mathrm{d}\mathbf{x} = \frac{\mathrm{d}\mathbf{x}}{\mathrm{d}\sigma}\mathrm{d}\sigma$$

is of course a vector, while $\mathrm{d}\sigma$ is a scalar. So their quotient is a vector:

$$\text{tangent vector to the curve} = \dot{\mathbf{x}} \equiv \frac{\mathrm{d}\mathbf{x}}{\mathrm{d}\sigma}$$

with contravariant components

$$\dot{x}^{\mu} \equiv \frac{\mathrm{d}x^{\mu}}{\mathrm{d}\sigma}.$$

(We shall always use the 'dot' notation for the derivative with respect to a parameter on a curve.)

How long is such a curve? The length of a short arc $\mathrm{d}\mathbf{x}$ at P involves the metric at P, and is

$$\mathrm{d}s = \sqrt{\{g_{\mu\nu}(\mathrm{P})\,\mathrm{d}x^{\mu}\,\mathrm{d}x^{\nu}\}}.$$

Integrating,

$$S = \text{length of curve between } \sigma_1 \text{ and } \sigma_2 = \int_{\sigma_1}^{\sigma_2} \sqrt{(g_{\mu\nu}\,\mathrm{d}x^{\mu}\,\mathrm{d}x^{\nu})}$$

$$= \int_{\sigma_1}^{\sigma_2} \sqrt{L}\,\mathrm{d}\sigma$$

after a simple change of variable, where L is the **Lagrangian**

$$L(x^{\mu}, \dot{x}^{\mu}) = g_{\mu\nu}(\mathbf{x})\dot{x}^{\mu}\dot{x}^{\nu}.$$

If the curve is to be a geodesic, its length is to be stationary against any small

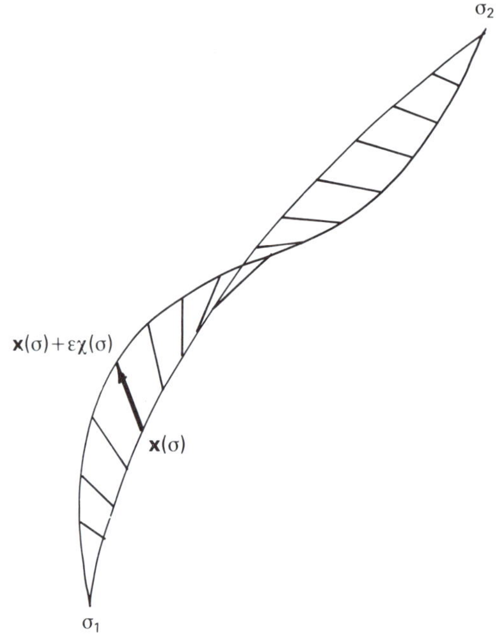

Fig. 3.2

alteration. Therefore envisage a slightly distorted curve (Fig. 3.2)

$$\mathbf{x}(\sigma) + \varepsilon\boldsymbol{\chi}(\sigma),$$

where ε is small, and the functions $\boldsymbol{\chi}$ specify the distortion. The ends of the curve are to remain fixed, so that $\boldsymbol{\chi}(\sigma_1) = 0$ and $\boldsymbol{\chi}(\sigma_2) = 0$. If $S(\varepsilon)$ is the length of the new curve, we require, for arbitrary distortion $\boldsymbol{\chi}$,

$$\left.\frac{\mathrm{d}S(\varepsilon)}{\mathrm{d}\varepsilon}\right|_{\varepsilon=0} = 0.$$

Now

$$S(\varepsilon) = \int_{\sigma_1}^{\sigma_2} \sqrt{[L(\mathbf{x} + \varepsilon\boldsymbol{\chi}, \dot{\mathbf{x}} + \varepsilon\dot{\boldsymbol{\chi}})]}\,\mathrm{d}\sigma,$$

giving

$$\frac{\mathrm{d}S(\varepsilon)}{\mathrm{d}\varepsilon} = \int_{\sigma_1}^{\sigma_2} \frac{1}{2\sqrt{L}}\left(\frac{\partial L}{\partial \dot{x}^\mu}\dot{\chi}^\mu + \frac{\partial L}{\partial x^\mu}\chi^\mu\right)\mathrm{d}\sigma \quad \text{(when taken at } \varepsilon = 0\text{)}$$

$$= \frac{1}{2}\int_{\sigma_1}^{\sigma_2}\left\{\frac{\mathrm{d}}{\mathrm{d}\sigma}\left(-\frac{1}{\sqrt{L}}\frac{\partial L}{\partial \dot{x}^\mu}\right) + \frac{1}{\sqrt{L}}\frac{\partial L}{\partial x^\mu}\right\}\chi^\mu(\sigma)\,\mathrm{d}\sigma$$

after integrating certain of the terms by parts, and using $\boldsymbol{\chi}(\sigma_1) = \boldsymbol{\chi}(\sigma_2) = 0$ in each case. If this is to vanish for arbitrary $\boldsymbol{\chi}$, then all of the equations

$$\frac{\mathrm{d}}{\mathrm{d}\sigma}\left(\frac{1}{\sqrt{L}}\frac{\partial L}{\partial \dot{x}^\mu}\right) - \frac{1}{\sqrt{L}}\frac{\partial L}{\partial x^\mu} = 0, \quad \mu = 1, 2, \ldots, n,$$

must be simultaneously satisfied along the curve.

Since only $n-1$ equations are needed to specify a curve in nD, the given equations cannot be independent. This reflects the fact that there are many ways to parametrize the same curve.

3.7 Affine parameters

In practice, we always choose σ to be an **affine parameter**, that is, one for which L is constant along the curve. This is always possible: for example, why not take $\sigma = s$, the length measured along the curve? This guarantees $L = 1$ throughout.

Such a choice has two advantages. First, the geodesic equations simplify to the *affine* or **Euler** form

$$\frac{\mathrm{d}}{\mathrm{d}\sigma}\left(\frac{\partial L}{\partial \dot{x}^\mu}\right) - \frac{\partial L}{\partial x^\mu} = 0, \quad \mu = 1, 2, \ldots, n; \tag{3.6}$$

and secondly, $L =$ constant is always available as a **first integral** of these equations and is a substantial aid in their solution.

In $(n+0)$D space, L is necessarily positive along a curve. But in the $(1+3)$D space of General Relativity, L may have either sign, or may even be zero. Thus we need to distinguish three different types of geodesic.

(i) **Timelike** $(L > 0)$. The tangent vector has positive squared-length everywhere along the curve, and may therefore be taken as a velocity four-vector. The geodesic itself is then physically interpreted as a possible worldline for an inertial observer. The affine parameter is then the proper time along the worldline. The first integral is taken as $L = 1$ (or perhaps $L = c^2$, depending on units).

(ii) **Null** $(L = 0)$. The tangent vector has zero squared-length, and lies on the lightcone at every point. Such a geodesic is appropriate as the worldline of a radar pulse, or of a massless 'particle' such as a photon or neutrino. The first integral $L = 0$ shows that the idea of proper time along the worldline is meaningless; though other affine parameters exist, they are not generally useful.

(iii) **Spacelike** $(L < 0)$. Such geodesics lie outside the lightcones at all points. They have mathematical interest only, from a physical point of view, since a spacelike geodesic can never be a worldline of anything.

Since $L =$ constant is a guaranteed integral of the Euler equations, a geodesic cannot change type along its length.

3.8 First integrals of the equations

In addition to $L =$ constant, there may be other first integrals of the Euler equations. Generally the metric $\mathbf{g}(\mathrm{P})$ is a function of the position P. In particular applications the spacetime may possess some kind of *symmetry* which allows us to choose coordinates in which $\mathbf{g}$ is independent of one of those coordinates, u, say. The Euler equation corresponding to the coordinate u then simplifies to

$$\frac{\mathrm{d}}{\mathrm{d}s}\left(\frac{\partial L}{\partial \dot{u}}\right)=0, \tag{3.7}$$

which may be obviously integrated to

$$\frac{\partial L}{\partial \dot{u}} = \text{constant along a geodesic.} \tag{3.8}$$

Extra first integrals of this kind are very useful for finding geodesics in specific applications.

A coordinate u of this kind is sometimes known as **ignorable**. The corresponding first integral is a **momentum**, in a generalized sense. Depending on the application, such 'momenta' may appear in a variety of guises: linear momentum, angular momentum, energy, and so on.

3.9 An example

It is crucially important to be able to handle geodesics. The worldlines of inertial observers are geodesics. An understanding of curvature depends heavily on the idea of a geodesic. Geodesics will be used to probe the darker recesses of black holes which might otherwise be missed. For such reasons we shall need to develop some techniques.

An easily understood, yet not too trivial, example is that of a geodesic on the surface of the unit sphere. Of course, we know the answer: the geodesics are the so-called great circles; but it is instructive to derive this result from the Lagrangian.

The metric on the surface of the unit sphere is

$$\mathrm{d}s^2 = \mathrm{d}\theta^2 + \sin^2\theta\,\mathrm{d}\phi^2,$$

where the coordinates are the customary colatitude θ and longitude ϕ; see section 3.1. The Lagrangian is thus

$$L = \dot{\theta}^2 + \sin^2\theta\dot{\phi}^2.$$

This provides us with an immediate first integral, which we take as $L = 1$, ensuring that the affine parameter on the geodesic is distance s measured on the surface of the sphere. Notice that the coordinate ϕ does not appear explicitly in L, and is thus 'ignorable': this yields another first integral

$$\frac{\partial L}{\partial \dot{\phi}} = \text{constant},$$

that is

$$\dot{\phi}\sin^2\theta = J \text{ (constant).} \tag{3.9}$$

Now eliminate $\dot{\phi}$ from these first integrals to obtain

$$\dot{\theta}^2 = 1 - \frac{J^2}{\sin^2\theta}.$$

This is a first-order ordinary differential equation for $\theta(s)$ as a function of s. By the way, it provides us with the geometrical significance of the constant J; since the left side is a square, the right side is obliged to be non-negative, and thus $\sin\theta > |J|$ throughout the motion. Therefore the colatitude is bounded by the values θ_0 and $\pi - \theta_0$, where the alternative constant θ_0 is fixed by $\sin\theta_0 = |J|$. (The sign of J is determined by the sense of the rotation.) It is usual to find that the constants of the motion are associated with the extremities of the motion in this way.

The variables are 'separable', in the sense that

$$\frac{\sin\theta\,\mathrm{d}\theta}{\sqrt{[\sin^2\theta - \sin^2\theta_0]}} = \mathrm{d}s.$$

Integration by straightforward standard methods gives

$$\cos\theta(s) = \cos\theta_0 \sin(s - s_0), \tag{3.10}$$

s_0 being the constant of integration. This constant merely fixes the origin of the affine parameter, and is usually dropped forthwith. If this is done, then s is measured from a crossing of the Equator $\theta = \pi/2$.

It remains to determine ϕ as a function of s. Eliminate θ between (3.9) and (3.10) to obtain

$$\dot{\phi}[1 - \cos^2\theta_0 \sin^2 s] = J.$$

This differential equation is separable:

$$\mathrm{d}\phi = \frac{J\,\mathrm{d}s}{1 - \cos^2\theta_0 \sin^2 s}.$$

Integration gives

$$\tan(\phi - \phi_0) = \sin\theta_0 \tan s, \tag{3.11}$$

where ϕ_0 is the longitude of the equatorial crossing. The equations (3.10) and (3.11), taken together, provide the parametric representation of a typical geodesic. Any such geodesic may be obtained by assigning suitable values to the disposable constants θ_0 and ϕ_0.

Of course, nobody would dream of dealing with great circles on the sphere in this way, except to illustrate techniques.

3—Notes and problems

1. The form of the metric contains information about the intrinsic curvature of the space. This information is usually well hidden, and it is not possible at a glance at the metric to say what the curvature happens to be. For example, of the following six metrics, which represent flat surfaces?

   (i) $\mathrm{d}r^2 + r^2\,\mathrm{d}\phi^2$
   (ii) $\mathrm{d}\theta^2 + \sin^2\theta\,\mathrm{d}\phi^2$
   (iii) $(\mathrm{d}u^2 + \mathrm{d}v^2)\exp(u - v)$
   (iv) $\dfrac{\mathrm{d}r^2}{1 - r^2} + r^2\,\mathrm{d}\phi^2$

(v) $\dfrac{dx^2 + dy^2}{\sqrt{(x^2 + y^2)}}$
(vi) $dw^2 + d\phi^2 \exp(-2w)$.

In fact, (i), (iii), and (v) are flat; (ii) and (iv) have constant positive curvature, while (vi) has constant negative curvature. The precise meaning of these statements—and their proof—will have to wait till Chapter 7.

2. **Changing coordinates** It is traditional to associate the axis of symmetry of spherical polar coordinates $r\theta\phi$ with the z-axis of Cartesian coordinates xyz. Of course, this is not at all necessary. For example, we may envisage polar coordinates $r\alpha\beta$ whose axis of symmetry is the x-axis. The question is: how are these two sets of polar coordinates related?

   Clearly the coordinate r has the same significance for each. The relation between the angles is easily found by comparing the two forms for the same typical unit vector

$$\begin{aligned}\mathbf{u} &= (\sin\theta\cos\phi, \sin\theta\sin\phi, \cos\theta)\\ &= (\cos\alpha, \sin\alpha\cos\beta, \sin\alpha\sin\beta).\end{aligned}$$

   Hence

$$\cos\theta = \sin\alpha\sin\beta \quad \text{and} \quad \tan\phi = \tan\alpha\cos\beta$$

   gives the 'old' $\theta\phi$ in terms of the 'new' $\alpha\beta$; this is usually the best way round.

   Show that

$$-\sin\theta\, d\theta = \cos\alpha\sin\beta\, d\alpha + \sin\alpha\cos\beta\, d\beta$$

   and

$$\sec^2\phi\, d\phi = \sec^2\alpha\cos\beta\, d\alpha - \tan\alpha\sin\beta\, d\beta.$$

   Hence show that under this change of coordinates

$$d\theta^2 + \sin^2\theta\, d\phi^2 = d\alpha^2 + \sin^2\alpha\, d\beta^2.$$

   Give reasons for supposing you could have saved yourself the trouble.

3. **Fake boundaries** x and y are Cartesian coordinates on the plane, and u and v are new coordinates defined by

$$u = -x + \ln y \quad \text{and} \quad v = x + y^2/2.$$

   Show that

$$dx^2 + dy^2 \quad \text{becomes} \quad \frac{y^2\, du^2 + dv^2}{y^2 + 1}$$

   where y is now a function of $u + v$, defined intrinsically by

$$\tfrac{1}{2}y^2 + \ln y = u + v.$$

   On account of the logarithm, y can take only positive values, so that the uv-coordinate system covers the upper half-plane only.

   It can sometimes happen that we are presented with a metric with a natural boundary of such a kind, which has nothing to do with any boundary that the space itself may have. Circumventing such limitations is important when we wish, for example, to study the *entire* physics of a black hole.

4. **Great circles on the unit sphere** When referred to the usual Cartesian axes, the position of the point with colatitude θ and longitude ϕ is given by

$$(x, y, z) = (\sin\theta\cos\phi,\ \sin\theta\sin\phi,\ \cos\theta).$$

The two orthogonal unit vectors

$$\mathbf{u}_1 = (1, 0, 0) \quad \text{and} \quad \mathbf{u}_2 = (0, \sin\theta_0, \cos\theta_0)$$

determine the plane of a great circle inclined at an angle θ_0 from the z-axis; the point 'at s' on this great circle is at

$$(x, y, z) = \mathbf{u}_1 \cos s + \mathbf{u}_2 \sin s = (\cos s, \sin\theta_0 \sin s, \cos\theta_0 \sin s).$$

Comparison reveals that

$$\cos\theta = \cos\theta_0 \sin s \quad \text{and} \quad \tan\phi = \sin\theta_0 \tan s.$$

This is exactly the parametric representation of (3.10) and (3.11). The moral is: Don't solve differential equations when you know the answer already.

Draw a diagram to clarify these matters.

5. **Symmetry** The unit sphere is spherically symmetric in an obvious way. The unit sphere *with added latitude and longitude coordinates* is axially symmetric only; this is why only ϕ is ignorable, giving

$$J_3 = \dot{\phi}\sin^2\theta,$$

constant along a great circle. We may take it as evident that on account of the symmetry there must be other such constants. Indeed, in the alternative coordinates α and β in Problem 2, β is ignorable, and there is another constant

$$J_1 = \dot{\beta}\sin^2\alpha.$$

Show that transforming to the original coordinates gives

$$J_1 = \dot{\theta}\sin\phi + \dot{\phi}\sin\theta\cos\theta\cos\phi.$$

Similarly, there is yet another constant

$$J_2 = \dot{\theta}\cos\phi - \dot{\phi}\sin\theta\cos\theta\sin\phi.$$

Show that

$$J_1{}^2 + J_2{}^2 + J_3{}^2 = L.$$

There is no system of coordinates in which all the first integrals can be written down by inspection. (For the *cognoscenti*: the reason lies with the group of rotations, which is not commutative.)

6. **The Hamiltonian recipe for a geodesic** Those who know about such things expect that where there is a Lagrangian, a Hamiltonian will not be far behind. Define the 'momentum'

$$p_\alpha = g_{\alpha\beta}\dot{x}^\beta;$$

this is nothing but the covariant version of the four-velocity $\dot{\mathbf{x}}$. Alternatively,

$$\dot{x}^\beta = g^{\beta\alpha}p_\alpha;$$

while the Euler equation may be written

$$\dot{p}_\mu = \frac{1}{2}\frac{\partial L}{\partial x^\mu}.$$

Show that the last two equations may be written

$$\dot{x}^\mu = \frac{\partial H}{\partial p_\mu} \quad \text{and} \quad \dot{p}_\mu = -\frac{\partial H}{\partial x^\mu}$$

where the **Hamiltonian** is

$$H(x, p) = \tfrac{1}{2} g^{\mu\nu}(P) p_\mu p_\nu.$$

(You will need to show that

$$\frac{\partial g^{\mu\nu}}{\partial x^\sigma} = -g^{\mu\alpha} g^{\nu\beta} \frac{\partial g_{\alpha\beta}}{\partial x^\sigma}.)$$

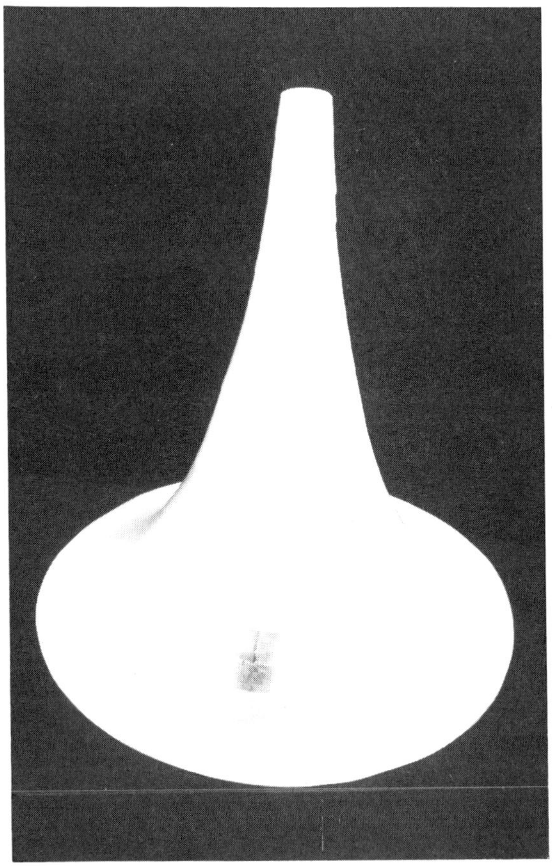

Fig. 3.3

7. For the metric

$$ds^2 = r^{-2}\,dr^2 + r^2\,d\phi^2$$

obtain the two first geodesic integrals

$$r^{-2}\dot{r}^2 + r^2\dot{\phi}^2 = 1 \quad \text{and} \quad r^2\dot{\phi} = a \text{ (constant)}.$$

Then obtain in sequence

$$\dot{r}^2 + a^2 = r^2,$$
$$r = a\cosh s,$$
$$\dot{\phi} = a^{-1}\operatorname{sech}^2 s,$$
$$\phi = \phi_0 + a^{-1}\tanh s.$$

The surface in question may be partly constructed in 3D, where it takes the form of a trumpet of infinite length with axial symmetry about the axis $r = 0$ (Fig. 3.3). This *model* cannot be continued beyond the rim at $r = 1$, but it is important to recognize that the *surface* itself is not to be thought of as ending there: it continues 'outward' for ever, as do the geodesics on it. None of the equations presented above shows any kind of difficulty at $r = 1$. There is no reason why they should, since the work is entirely independent of any imagined embedding in 3D, being 'intrinsic' from start to finish.

8. The **tractrix** (or **tractory**) is a plane curve with the property that the tangent at any point A meets a certain fixed line at B so that AB = 1; in Fig. 3.4, the fixed line is the z-axis. The distance of A from the axis is r, and w is the arc-length along the curve measured from its cusp. Fig. 3.4 reveals that, under a small movement dw along the curve,

$$dw:dr:dz = 1:-r:\sqrt{(1-r^2)}.$$

In particular, since $dw^2 = r^{-2}\,dr^2$, show that rotating the curve about the axis generates the trumpet-shaped surface of Problem 7.

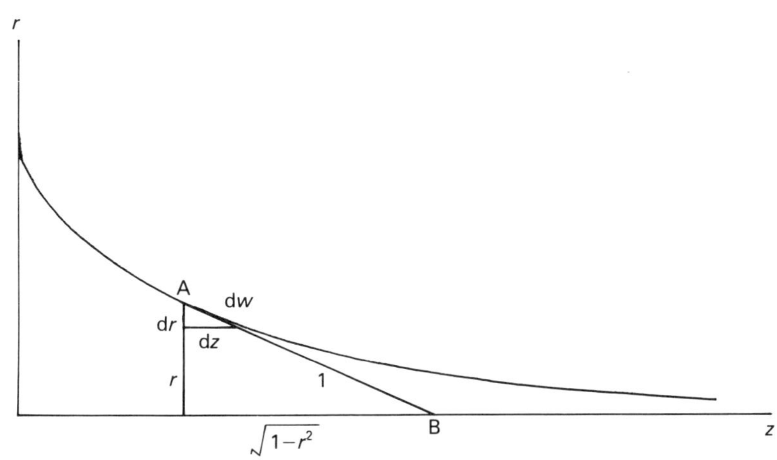

Fig. 3.4

Show that $r = \exp(-w)$, and that if coordinates w, ϕ are used, the metric takes the form given at Problem 1(vi).

Show incidentally that the tractrix is the path followed by a certain point on a trailer when the towing vehicle drives along a straight line (hence the name, of course).

9. The matrix $\mathbf{g}$ formed from the metric tensor is real and symmetric. Its eigenvalues therefore are all real, and its column eigenvectors $v_1, v_2, \ldots$ are (or, in case of degeneracy, may be taken to be) mutually orthogonal in pairs. Regard the elements of each eigenvector v_i as the contravariant components of a *geometrical* vector $\mathbf{V}_i$. Show that these geometrical vectors are orthogonal in pairs, and that the number n_1 of these vectors with positive squared-length is precisely the number of positive eigenvalues of $\mathbf{g}$. (Consider $v^{\mathrm{T}}{}_i \mathbf{g} v_j$ for various i, j.)

   The vectors obtained in this way have a negligible geometrical significance. They are useful only for establishing the connection between signature and the signs of the eigenvalues of $\mathbf{g}$.

10. Show that the metric

$$ds^2 = du\,dv + du\,dw + du\,dz + dv\,dw + dv\,dz + dw\,dz$$

belongs to a space of signature (1 + 3)D.

4

Spacetime near a massive sphere

This chapter deals with a particularly important special case of a curved spacetime, that due to the presence of a spherically symmetric distribution of mass. The metric satisfies the Einstein field equations for empty spacetime, and we shall not be ready for its derivation until Chapter 8. For the moment, we shall take the metric as given, and consider its consequences, just as we might for Newtonian gravity where we are prepared to accept the inverse-square law before solving Poisson's equation for the gravitational potential.

Our chief interest is in the evidences in favour of Einstein's theory to be found within the confines of the Solar System. Here we have a very prominent, very nearly spherical mass—the Sun—whose gravitational environment is continuously monitored, as it were, by a host of test particles in free fall (the planets and many artificial probes). Additionally, the accuracy of astronomical observations has reached an impressively high level, high enough to provide crucial tests for the adequacy of the Einstein view of gravity. One or two of the more recent tests are Earth-bound, and use the Earth as the relevant spherical mass.

The approach will be one of first-order approximation. We have already seen that the scale of curvature required to produce 'ordinary' gravity is very small, and it turns out to be enough to calculate the first-order differences between the Newtonian and the Einstein theories. In all cases, the Einstein theory will appear as the better of the two.

4.1 Polar coordinates in Minkowski space

It is usually convenient to express spherically symmetric problems in spherical polar coordinates. As a preliminary we transform the Minkowski metric in this way.

The metric in question is

$$c^2\,\mathrm{d}\tau^2 = c^2\,\mathrm{d}t^2 - \mathrm{d}x^2 - \mathrm{d}y^2 - \mathrm{d}z^2. \tag{4.1}$$

Under the expected transformation,

$$x = r\sin\theta\cos\phi, \quad y = r\sin\theta\sin\phi, \quad z = r\cos\theta, \quad t = t, \tag{4.2}$$

the metric becomes

$$c^2\,\mathrm{d}\tau^2 = c^2\,\mathrm{d}t^2 - \mathrm{d}r^2 - r^2\,\mathrm{d}\theta^2 - r^2\sin^2\theta\,\mathrm{d}\phi^2. \tag{4.3}$$

This system of coordinates lays special emphasis on the t-axis: it is the worldline $r = 0$ of the centre of the system, and is destined to be the worldline of the centre of the massive sphere. The coordinate r measures radial distance in the surface $t = \text{constant}$; however, it is better for our purposes to think of it as labelling the sphere of circumference $2\pi r$ and area πr^2, since the radial coordinate will keep this meaning in the curved spacetime soon to be introduced. The angular coordinates θ and ϕ have the usual significance of colatitude and longitude.

The geodesics are still straight lines, of course, and their form in the present coordinates is easily found by the obvious substitution. Alternatively, we may solve for the geodesics directly. Without loss of generality, the spherical symmetry allows us to consider equatorial geodesics alone, along which $\theta = \pi/2$, and for which the Lagrangian integral simplifies to

$$L = c^2\dot{t}^2 - \dot{r}^2 - r^2\dot{\phi}^2 = c^2.$$

In addition, the coordinates t and ϕ are both ignorable, and we have two further integrals of the motion

$$\dot{t} = \gamma \qquad \text{(energy constant)},$$
$$r^2\dot{\phi} = J \qquad \text{(angular momentum constant)}.$$

If we eliminate $\dot{t}$ and $\dot{\phi}$ from the three integrals, we are left with

$$\dot{r}^2 + \frac{J^2}{r^2} = c^2(\gamma^2 - 1),$$

which suggests, after multiplication by $m/2$,

radial kinetic energy + potential energy for centrifugal force = total energy,

thus justifying the names given to the constants γ and J.

It is always useful to emphasize those values of r for which $\dot{r} = 0$; these are the distances of closest approach to or furthest removal from the centre (**pericentre, apcentre**—in appropriate cases **perihelion, perigee, periastron**, and **aphelion, apogee, apastron**). For the above equation there is just one (positive) r_0 for which $\dot{r} = 0$, namely

$$r_0 = \frac{J}{c}(\gamma^2 - 1)^{-1/2}.$$

The previous equation may now be written

$$\dot{r}^2 = c^2(\gamma^2 - 1)\frac{r^2 - r_0^{\,2}}{r^2}.$$

Since the right side is positive for $r > r_0, r_0$ is a pericentre (not an apcentre).

The solution is straightforward, and gives part of the parametric representation of a typical equatorial geodesic:

$$r^2 = r_0{}^2 + c^2\tau^2(\gamma^2 - 1). \tag{4.4a}$$

The remainder of the representation is to be found by integrating the earlier equations to obtain t and ϕ:

$$t = t_0 + \gamma\tau, \tag{4.4b}$$

$$\phi = \phi_0 + \arctan\frac{c\tau}{r_0}\sqrt{[\gamma^2 - 1]}. \tag{4.4c}$$

And, of course,

$$\theta = \pi/2, \quad \text{constant at all times,} \tag{4.4d}$$

since we have restricted the discussion to geodesics in the equatorial plane. As τ goes from $-\infty$ to $+\infty$, the whole geodesic is traversed once, as specified by (4.4a)–(4.4d); the pericentre is encountered at $\tau = 0$.

4.2 The Schwarzschild metric

In 1916, Karl Schwarzschild obtained the first exact nontrivial solution of the Einstein field equations for gravity in empty space. In one coordinate system, it is

$$c^2\,\mathrm{d}\tau^2 = c^2(1 - a/r)\,\mathrm{d}t^2 - \frac{\mathrm{d}r^2}{1 - a/r} - r^2\,\mathrm{d}\theta^2 - r^2\sin^2\theta\,\mathrm{d}\phi^2. \tag{4.5}$$

There is one disposable parameter, namely a. When $a = 0$, the Minkowski metric in polar coordinates is recovered. If a is not zero, the metric is curved, but still spherically symmetric. The way in which θ and ϕ enter in combination with r, namely

$$r^2(\mathrm{d}\theta^2 + \sin^2\theta\,\mathrm{d}\phi^2),$$

reveals that r still labels a sphere of area πr^2, and that θ and ϕ are still colatitude and longitude on this sphere. However, radial distances are different, and timekeeping now depends on the value of r.

As an example, let us examine the matter of radial distance. To be precise, let us ask for the distance s between the two events (t, r_1, θ, ϕ) and (t, r_2, θ, ϕ). Since t, θ, ϕ are the same for the two events, the relevant residue of the metric is

$$\mathrm{d}s^2 = \frac{\mathrm{d}r^2}{1 - a/r},$$

which immediately leads to

$$\text{distance between } r_1 \text{ and } r_2 = \int_{r_1}^{r_2} \frac{\mathrm{d}r}{\sqrt{(1 - a/r)}}.$$

If we suppose that a is very small, we may write

$$\text{distance} = \int_{r_1}^{r_2} \mathrm{d}r\{1 + a/2r + O(a^2)\} = r_2 - r_1 + \tfrac{1}{2}a\ln\frac{r_2}{r_1}. \tag{4.6}$$

This reveals the extent to which the radial coordinate fails to represent radial distance: the discrepancy is logarithmic.

It will transpire that a is positive, for physical reasons. It clearly follows that the metric is in some sense unsatisfactory at $r = a$, and we shall have to consider this further under the heading of 'black holes' (Chapter 9). Meanwhile, it is enough to note that a is expected to be very small compared to typical distances in the Solar System; in particular, a will be very much less than the radius of the massive sphere responsible for the spacetime curvature. For all accessible values of r, therefore, the metric is perfectly well-behaved. (Inside the sphere, of course, the metric takes a different 'interior' form: it satisfies the non-vacuum field equations and, for an object like our Sun, is well-behaved everywhere. We shall return to this in Chapter 12.)

4.3 Radial acceleration and the meaning of the constant *a*

In order to understand the physical significance of the constant a we shall need to consider mechanics in the Schwarzschild environment; let us consider free fall in the radial direction (that is, with θ and ϕ both constant throughout). In this case the Lagrangian integral, which guarantees that the proper time τ is used as the parameter along the worldline, reduces to

$$L = c^2(1 - a/r)\dot{t}^2 - \frac{\dot{r}^2}{1 - a/r} = c^2 \quad (\text{parameter} = \tau).$$

Since t is ignorable, we have also

$$(1 - a/r)\dot{t} = \gamma \quad (\text{'energy'}).$$

Eliminating t and rearranging yields

$$c^2\gamma^2 - \dot{r}^2 = c^2(1 - a/r). \tag{4.7a}$$

Differentiating with respect to τ now gives

$$\ddot{r} = -\frac{ac^2}{2r^2} \tag{4.7b}$$

as the equation of motion of a small point-mass in radial free fall.

Now compare this with the corresponding Newtonian equation for a point-mass μ in radial free fall near a spherical mass M centred on the origin:

$$\mu\frac{\mathrm{d}^2r}{\mathrm{d}t^2} = -\frac{GM\mu}{r^2} \quad (\text{Newtonian})$$

where G is the **gravitational constant**. In view of the fact that for slow motions the differentiations with respect to t and to τ are virtually indistinguishable, we can make the immediate identification

$$ac^2 = 2GM. \tag{4.8}$$

Thus the constant $a/2$ is nothing more or less than the **mass** of the sphere, expressed in unfamiliar units.

With the values

$$G = 6.67 \times 10^{-11}\,\mathrm{N\,m^2\,kg^{-2}}$$
$$c = 2.998 \times 10^8\,\mathrm{m\,s^{-1}}$$

and the known masses and radii of the Sun and the Earth, we have

	Sun	Earth
Mass:	1.99×10^{30} kg	5.98×10^{24} kg
a:	2.96 km	8.92 mm
Radius (R):	6.96×10^{5} km	6378 km.

It is evident that the ratio a/r can never exceed a maximum a/R. For the Sun as centre this maximum is about 4×10^{-6}; for the Earth, it is even less, 1.4×10^{-9}. The approximations of this chapter are fully justified by the smallness of these numbers.

4.4 Timekeeping in an Earth-bound laboratory

Imagine that we are in a laboratory on the Equator of the Earth. Our worldline is a helix in the spacetime plot, on account of the rotation of the Earth (we shall ignore the Earth's motion in its orbit). This worldline is given parametrically by

$$
\begin{aligned}
t &= \gamma\tau && (\gamma \text{ constant: steady timekeeping})\\
r &= \text{constant} && (\text{the Earth radius})\\
\theta &= \pi/2 && (\text{on the Equator})\\
\phi &= \omega\tau && (\omega \text{ constant: steady rotation}).
\end{aligned}
$$

(Of course, this worldline is *not* a geodesic, since the laboratory is certainly not in free fall, and therefore certainly not inertial.) The contravariant velocity-vector on this worldline has constant components,

$$\dot{x}^{\mu} = (\gamma, 0, 0, \omega).$$

The use of τ to parametrize the curve entails a relation between γ and ω,

$$g_{\mu\nu}\dot{x}^{\mu}\dot{x}^{\nu} \equiv c^2\left(1 - \frac{a}{r}\right)\gamma^2 - r^2\omega^2 = c^2.$$

Physically, this ensures that τ is the proper time in the laboratory.

A stationary observer a long way off will measure the velocity of the laboratory as

$$v = r\frac{\mathrm{d}\phi}{\mathrm{d}t} = \frac{r\omega}{\gamma}.$$

Eliminating ω from the last two equations, and rearranging, gives

$$\gamma = \left[1 - \left(\frac{v^2}{c^2} + \frac{a}{r}\right)\right]^{-1/2}.$$

Thus the distant observer will see the Earth-bound clock as running slow by the factor γ, in which the familiar effect of velocity is now inseparably bound up with the gravitational potential term a/r.

4.5 The Pound–Rebka–Snider experiment

Since the timekeeping of a clock is predicted to depend on its position and motion, neighbouring clocks need not be expected to keep quite the same time. The discrepancy is given by change in γ on moving through the displacement **h** from one clock to the other; to sufficient accuracy,

$$\frac{\tau_2}{\tau_1} = 1 - c^{-2}\mathbf{h}\cdot\nabla\left(\tfrac{1}{2}v^2 + \frac{ac^2}{2r}\right).$$

The brackets on the right enclose the sum of the centrifugal potential $v^2/2$ and the gravitational potential $ac^2/2r$, each for unit mass. The gradient of the total potential is simply $-\mathbf{g}$, the apparent gravitational acceleration local to the laboratory. ('Apparent', since the centrifugal acceleration is necessarily included. The way in which the two kinds of acceleration inevitably appear together provides the theoretical justification of the null result of the Eötvös experiment: within the Einstein framework, the null result is obligatory.) Finally, the timekeeping discrepancy is the ratio

$$\frac{\tau_2}{\tau_1} = 1 + \frac{\mathbf{h}\cdot\mathbf{g}}{c^2} = 1 + \frac{Hg}{c^2};$$

here H is the apparent vertical separation of the clocks.

In 1960, Pound and Rebka compared timekeeping at different levels in the laboratory; an improved version of the experiment was carried out by Pound and Snider in 1965. As described in Section 1.8, the technique depends on an application of the Mössbauer effect. The Mössbauer emitter was mounted above the absorber (a vertical separation H of about 22 m was used in the experiments), and the shift in the absorption peak was measured. The results agreed well with the relation

$$\delta\nu/\nu = Hg/c^2,$$

thus verifying the presence of the timekeeping discrepancy.

It would be wrong to regard this result as more than a rather weak confirmation of the theory; indeed, quantum mechanics suggests that any theory of gravity needs to provide for a frequency shift of this kind. Without such a shift, perpetual motion becomes conceivable: use an energy mc^2 to manufacture a particle of mass m, and drop it through a height H; convert the now available energy $mc^2 + mgH$ to a single photon, and direct it back to the original point of departure. Unless the photon loses energy—and hence frequency—as it climbs, there is a net gain of energy. This seems contrived, but it is enough to be able to think about it to acknowledge that a gravitational frequency shift is to be expected.

More recently it has been possible to observe timekeeping discrepancies more directly, with the development of very accurate atomic clocks. One experiment has involved flying one of a pair of identical clocks round the world by commercial airlines, comparing it with its twin before and after. To carry out such a procedure entails a careful logging of altitude r and speed v, in order that γ should be correctly integrated over the flight. Easterly and westerly trips yield different results, on account of the rotation of the Earth. As expected, the theory is confirmed.

4.6 Orbits

A planet of the Solar System is in free fall in the gravitational field of the Sun, and the shape of a typical planetary orbit is a very powerful test of the adequacy of any theory of gravity.

An **orbit** is not a worldline, but rather the projection of the worldline on to a surface $t = \text{constant}$. Its shape is obtained by eliminating all reference to time (either t or τ) from the equations of motion.

Without any loss of generality, we restrict ourselves to equatorial orbits, for which $\theta = \pi/2$ at all times. The Lagrangian first integral, which guarantees that τ is the parameter along the worldline, simplifies to

$$L = c^2(1 - a/r)\dot{t}^2 - \frac{\dot{r}^2}{1 - a/r} - r^2\dot{\phi}^2 = c^2.$$

Additionally, we have the conserved quantities

$$(1 - a/r)\dot{t} = \gamma \quad \text{and} \quad r^2\dot{\phi} = J$$

involving two disposable constants γ and J. There is also the evident relation

$$\dot{r} = \frac{\mathrm{d}r}{\mathrm{d}\phi}\dot{\phi}.$$

Eliminating $\dot{t}, \dot{r}$ and $\dot{\phi}$ from the last four relations gives

$$\frac{J^2}{r^4}\left(\frac{\mathrm{d}r}{\mathrm{d}\phi}\right)^2 = c^2(\gamma^2 - 1) + \frac{ac^2}{r} - \frac{J^2}{r^2} + \frac{aJ^2}{r^3}.$$

Traditionally, the variable $u = 1/r$ is used in orbit problems, as it makes the work slightly simpler. With this change we have

$$\left(\frac{\mathrm{d}u}{\mathrm{d}\phi}\right)^2 = au^3 - u^2 + \alpha u + \beta \tag{4.9}$$

as a differential equation for the shape of the orbit. The constants γ and J have been exchanged for the alternative disposable constants α and β, defined by

$$\alpha = \frac{ac^2}{J^2} \quad \text{and} \quad \beta = \frac{c^2(\gamma^2 - 1)}{J^2}. \tag{4.10}$$

When $\gamma > 1$, the constant β has a simple geometrical significance (Problem 4).

The differential equation is easily dealt with by separation of variables, and the general solution is

$$\phi - \phi_0 = \int_{u_0}^{u} \frac{\mathrm{d}u}{\sqrt{[au^3 - u^2 + \alpha u + \beta]}} \tag{4.11}$$

where $u = u_0$, $\phi = \phi_0$ is one of the points on the orbit. This is a very simple result, and implies that u is an elliptic function of ϕ. Unfortunately, an understanding of elliptic functions is not commonly found these days, in spite of the fact that they turn up so frequently in ordinary mechanics and in other branches of physics. The first-order approximations to be used below avoid the explicit use of elliptic functions.

4.7 Planetary orbits

A planet lacks the energy to escape to infinity, and its orbit lies entirely between an aphelion $r_1 = 1/u_1$ and a perihelion $r_2 = 1/u_2$. At these values of r, the derivative

$du/d\phi = 0$; thus we may recast the right side of (4.9) as

$$au^3 - u^2 + \alpha u + \beta = (u - u_1)(u_2 - u)(Au + B), \tag{4.12}$$

in which the constants A and B are chosen to get the terms $au^3 - u^2$ exactly right. In fact, the right side has to be

$$(u - u_1)(u_2 - u)[1 - a(u_1 + u_2 + u)]$$

and the general solution may thus be approximated by

$$\phi - \phi_0 = \int_{u_0}^{u} \frac{du}{\sqrt{[(u - u_1)(u_2 - u)]}} [1 + \tfrac{1}{2}a(u_1 + u_2 + u) + O(a^2)]. \tag{4.13}$$

The square root may be rationalized with the useful substitution

$$u = \frac{u_1 + u_2 q^2}{1 + q^2};$$

see the comment in Problem 3. As q goes from $-\infty$ through 0 to $+\infty$, u goes from perihelion u_2 to aphelion u_1 and back to u_2. The overall change in ϕ between successive perihelia is thus calculated as

$$\Delta\phi = 2\pi + \tfrac{3}{2}\pi a(u_1 + u_2) + O(a^2) \tag{4.13a}$$

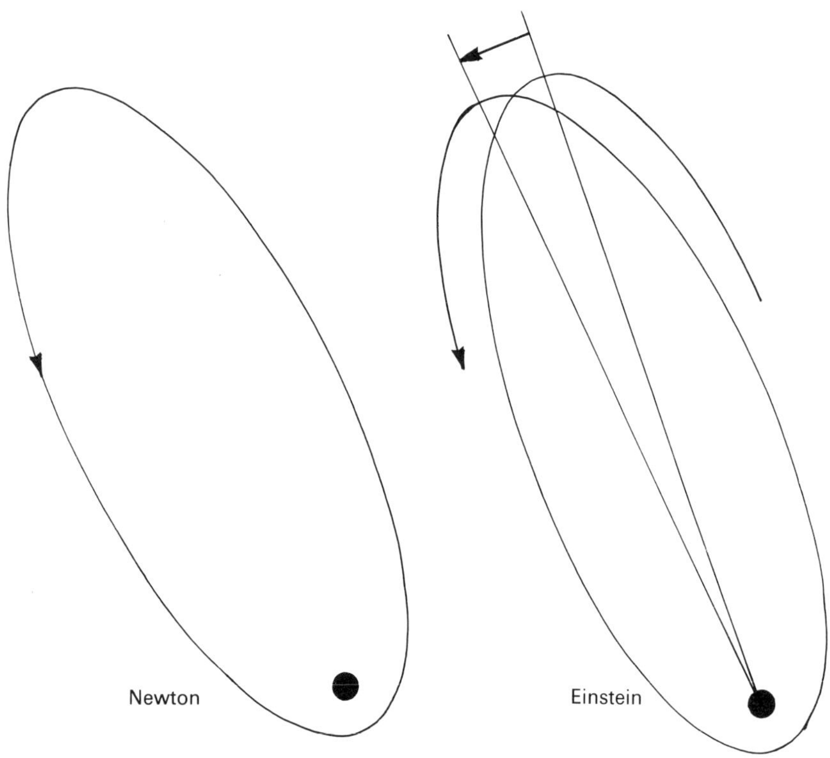

Fig. 4.1

after a standard integration (Problem 6). The corresponding result for Newtonian gravity is 2π exactly: in the Newtonian case, the orbit is **re-entrant**. Thus, in the Einstein case, there is an **anomalous perihelion precession**, in the same sense as that of the planet's motion, of amount $3\pi a/R$ radians per orbit, where R is an appropriate mean distance, $R = 2/(u_1 + u_2)$ (Fig. 4.1).

During the nineteenth century, the behaviour of the planet Mercury in its orbit round the Sun was puzzling. The orbit is not re-entrant, and the perihelion precession is quite large. Almost all of this precession was accounted for a resulting from the perturbing effect of the other planets, particularly Jupiter and Saturn. However, in 1845 Leverrier calculated an obstinate residue which it was certain could not be explained in this way. Several suggestions were made at the turn of the century: perhaps there was an undiscovered planet (provisionally named Vulcan) with an orbit inside that of Mercury; perhaps the Sun was more oblate than so far believed. Or perhaps—as James Jeans suggested—the Newtonian law of gravity required modification.

The unexplained residue was in fact slightly over 40 seconds of arc per century, the observational error being not more than about one second of arc. Triumphantly, Einstein's theory provided an explanation: with $R = 58 \times 10^6$ km and $a = 2.96$ km, the anomalous precession is about 3×10^{-6} radians per orbit, enough to abolish the discrepancy to within the observational error.

4.8 Deflection of light near the Sun

We have already seen that the Principle of Equivalence suggests that light ought to 'fall' towards a massive body. We are now in a position to perform a quantitative calculation of the effect.

A photon must follow a null geodesic, for which $\gamma = \infty$. The constant β must, however, remain finite, and thus $J = \infty$ and $\alpha = 0$ (see (4.10)). The equation (4.9) for the orbit—which is essentially the track followed by a ray of light—therefore reduces to

$$\left(\frac{\mathrm{d}u}{\mathrm{d}\phi}\right)^2 = au^3 - u^2 + \beta.$$

At perihelion $u_1, \mathrm{d}u/\mathrm{d}\phi = 0$, implying that $\beta = {u_1}^2 - a{u_1}^3$, and

$$\left(\frac{\mathrm{d}u}{\mathrm{d}\phi}\right)^2 = (u_1 - u)[(u_1 + u) - a({u_1}^2 + u_1 u + u^2)].$$

This leads to the approximation

$$\phi - \phi_0 = \int_{u_0}^{u} \frac{\mathrm{d}u}{\sqrt{({u_1}^2 - u^2)}}\left[1 + \tfrac{1}{2}a\frac{{u_1}^2 + u_1 u + u^2}{u_1 + u}\right] + O(a^2). \tag{4.14}$$

Here the appropriate substitution is $u = u_1(1 - q^2)/(1 + q^2)$; as q goes from -1 through 0 to $+1$, u goes from 0 to u_1 and back to 0, traversing the entire orbit once. The consequent change in ϕ is now found to be

$$\Delta\phi = \pi + 2au_1 \tag{4.15}$$

(Problem 8). If the light ray were straight, $\Delta\phi$ would be precisely π. A deflection of amount $2au_1$ is therefore predicted (Fig. 4.2).

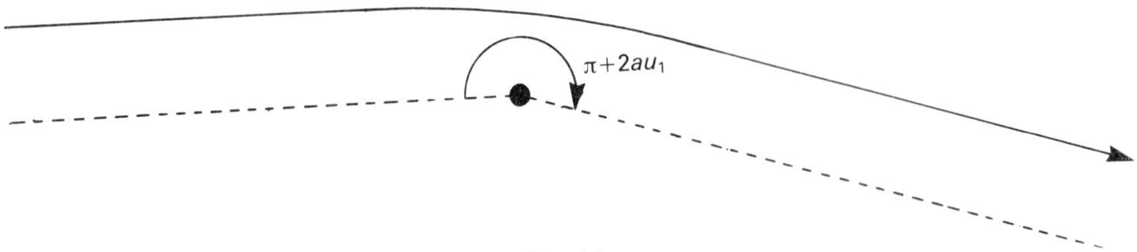

Fig. 4.2

In the case of a beam of light grazing the Sun, $r_1 = 6.96 \times 10^5$ km and $a = 2.96$ km. The deflection is therefore 8.51×10^{-6} radians, or just over 1.75 seconds of arc.

At the earliest possible opportunity after the First World War, Arthur Eddington organized expeditions to the track of a total Solar eclipse, in order to photograph the star field in the immediate neighbourhood of the Sun's disc. The atmospheric turbulence and sudden temperature changes which are characteristic of an eclipse made the observations difficult. Nevertheless, comparing the photographs with others taken at a different time of year made it clear that the deviation is present, and is of the right order. Eddington's original measurements have not been significantly improved on in the visible part of the spectrum, but more recently radio telescopes have been used by Shapiro to observe the occultation by the Sun of powerful radio sources, establishing beyond doubt the existence and the size of the effect.

4.9 Radar ranging

Let us now turn to the measurement of distances—in the Solar System, say—by radar methods. In contrast with the orbit problems examined so far, the theory involves the time in an essential manner. The principle is to measure the time interval (local proper time, of course) between the transmission and the later receipt of a pulse of electromagnetic radiation reflected from a distant target; the **radar distance** of the target is obtained in this way.

The pulse follows a null geodesic which needs to be determined. As before, and without loss of generality, we deal with trajectories in the equatorial plane. Once again, we have enough integrals to make headway:

$$c^2(1 - a/r)\dot{t}^2 - \frac{\dot{r}^2}{1 - a/r} - r^2\dot{\phi}^2 = 0 \quad \text{(null worldline)},$$

$$(1 - a/r)\dot{t} = A \quad \text{(constant)},$$

$$r^2\dot{\phi} = B \quad \text{(constant)}.$$

Additionally,

$$\dot{r} = \frac{\mathrm{d}r}{\mathrm{d}t}\dot{t}.$$

(Dot differentiations are with respect to some affine parameter σ or other; τ itself is not defined along a null geodesic. The constants A and B have a physical significance only through their ratio $\kappa = B/A$.) Elimination of $\dot{t}$, $\dot{r}$, and $\dot{\phi}$ leads to the single equation

$$\left(\frac{\mathrm{d}r}{\mathrm{d}t}\right)^2 = (1 - a/r)^2\left(c^2 - \kappa^2\frac{r-a}{r^3}\right). \tag{4.16}$$

As before, it is convenient to express the right side in terms of the perihelion r_1:

$$\frac{c^2}{\kappa^2} = \frac{r_1 - a}{{r_1}^3}.$$

The approach is as before: we work to first order in a, and make an appropriate rationalizing substitution, and integrate. The details are straightforward, and are relegated to Problem 9. The result is that the radar distance cT from perihelion to some other point with $r = R$ is

$$cT = \sqrt{(R^2 - {r_1}^2)} + a\left(\frac{q}{2} + \ln\frac{1+q}{1-q}\right) + O(a^2), \tag{4.17}$$

where $q^2 = (R - r_1)/(R + r_1)$. The first term is clearly the 'straight line' result, and the second is an increase arising from the presence of the Sun.

Shapiro's early radar experiments to check this result used the planet Venus as a target. The chief difficulty was that the reflected signal was extremely weak after a round trip of about five hundred million kilometres. In any case, Venus itself is a large rough object whose distance is therefore not very precise: the predicted effect is that, for a ray grazing the edge of the Sun, the radar diameter of the orbit of Venus is apparently increased by a mere 36.8 km. The corresponding extra radar delay is about 250 microseconds.

More recently, transponding space probes have made the observations much easier, and the existence of the effect is now beyond doubt.

4—Notes and problems

1. Verify that under the transformation

$$x = r\sin\theta\cos\phi, \quad y = r\sin\theta\sin\phi, \quad z = r\cos\theta,$$

the metric

$$\mathrm{d}x^2 + \mathrm{d}y^2 + \mathrm{d}z^2$$

goes over to

$$\mathrm{d}r^2 + r^2\,\mathrm{d}\theta^2 + r^2\sin^2\theta\,\mathrm{d}\phi^2.$$

Relate this result to equations (4.1), (4.2), and (4.3).

2. The parametric form for a straight line in the usual Minkowski coordinates t and $\mathbf{r} \equiv (x, y, z)$ is

$$t = t_0 + \gamma\tau,$$
$$\mathbf{r} = \mathbf{r}_0 + \gamma\mathbf{V}\tau,$$

where $t_0, \mathbf{r}_0$ and $\mathbf{V}$ are constants, and

$$\gamma = (1 - \mathbf{V}^2/c^2)^{-1/2}.$$

By transforming coordinates, obtain the parametric form in spherical polars $(tr\theta\phi)$. How is the choice of constants to be made to recover the parametric representation of (4.4a)–(4.4d)?

3. **Radial distance** The relation between the coordinate r and actual radial distance obtained in the text is approximate. However, the exact relation is readily found.

We need to evaluate

$$\text{distance between } r_1 \text{ and } r_2 = \int_{r_1}^{r_2} \frac{\mathrm{d}r}{\sqrt{([1-a/r])}}.$$

Show that the substitution $r = a/(1-q^2)$ rationalizes the square root, and leads to the indefinite integral

$$I = \frac{a}{2}\ln\frac{1+q}{1-q} + \frac{aq}{1-q^2}.$$

This is the exact result. Show that

$$I = r + \frac{a}{2}\ln r + \text{constant} + O(a^2)$$

and hence recover the result of (4.6).

We shall repeatedly need to evaluate integrals whose integrands involve a square root of a quadratic expression. It is fashionable to use a trigonometric substitution to dispose of the square root. However, I would recommend an *algebraic* substitution in all such cases, based on the 'Pythagoras identity'

$$(1+q^2)^2 \equiv (1-q^2)^2 + (2q)^2.$$

The integrand is then a rational function of q, and may be dealt with by the usual partial fraction techniques.

4. When an orbit possesses an asymptote as r goes to infinity, the **impact parameter** b is defined as the perpendicular distance of the asymptote from the origin. Fig. 4.3 shows that, as $r \to \infty$, $bu \equiv b/r \to \sin\phi$, and hence

$$b\frac{\mathrm{d}u}{\mathrm{d}\phi} \to 1 \quad \text{as } u \to 0.$$

Show that, for the equation (4.9), this implies that $\beta = 1/b^2$.

5. **Elliptic functions** appear regularly, even in the simplest examples of Newtonian mechanics. A simple pendulum of length a and substantial angle of swing θ is governed by the energy equation

$$\tfrac{1}{2}a^2\dot{\theta}^2 - ga\cos\theta = -ga\cos\theta_0 \quad \text{(constant)},$$

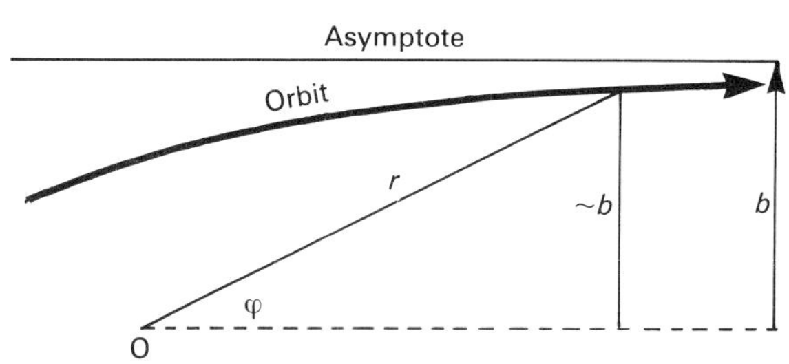

Fig. 4.3

where θ_0 is the maximum angle of swing. The substitution $z = \cos\theta$ now leads to

$$\dot{z}^2 = \frac{2q}{a}(z - z_0)(1 - z^2).$$

The right side is a *cubic polynomial* in z; it follows that the solution is expected to include an elliptic function. Special values of z_0 (namely ± 1) lead to something simpler; investigate them.

It is curious that functions of such importance should now receive so little attention.

6. **Perihelion precession** Under the substitution

$$u = \frac{u_1 + u_2 q^2}{1 + q^2}$$

show that

$$\int_{u_0}^{u} \frac{du}{\sqrt{[(u - u_1)(u_2 - u)]}}[1 + \tfrac{1}{2}a(u_1 + u_2 + u)]$$

transforms to

$$\int_{q_0}^{q} \frac{2\,dq}{1 + q^2}\left(1 + \tfrac{1}{2}a(u_1 + u_2) + \tfrac{1}{2}a\frac{u_1 + u_2 q^2}{1 + q^2}\right).$$

Evaluate this integral over the range $-\infty$ to $+\infty$, to obtain (4.13a).

7. Show that the perihelion precession of a geosynchronous Earth satellite is nearly 2×10^{-9} radians per orbit. (Ignore the gravity field of the Sun. Why? It turns out that the Sun exerts a small rotational *geodesic effect* on the entire Earth–Satellite system; see Chapter 5.)

8. **Deflection of light.** Using the substitution $u = u_1(1 - q^2)/(1 + q^2)$ in (4.14), evaluate over one traverse of the orbit

$$\int \frac{du}{\sqrt{(u_1{}^2 - u^2)}}\left[1 + \tfrac{1}{2}a\frac{u_1{}^2 + u_1 u + u^2}{u_1 + u}\right]$$
$$= \int_{-1}^{+1} \frac{2\,dq}{1 + q^2}\left(1 + au_1\frac{3 + q^4}{4(1 + q^2)}\right)$$
$$= \pi + 2au_1.$$

9. **Radar ranging** Show that

$$\left(1 - \frac{a}{r}\right)^{-1}\left(1 - \frac{r_1{}^3}{r^3}\frac{r - a}{r_1 - a}\right)^{-1/2}$$
$$= \frac{r}{\sqrt{(r^2 - r_1{}^2)}} + \frac{a}{2}\frac{1}{\sqrt{(r^2 - r_1{}^2)}}\frac{2r + 3r_1}{r + r_1} + O(a^2).$$

Use the substitution

$$r = r_1\frac{1 + q^2}{1 - q^2}$$

to obtain the result of the text.

5

Parallel transport and the principle of equivalence

5.1 How can the meaning of 'parallel' be saved?

Vectors have length and direction. Since length is a scalar, there is no problem in comparing the lengths of two vectors at different places. On the other hand, unless spacetime is flat, it is not possible to compare their directions in any unambiguous way, since 'parallel' loses its meaning over an extended region. The best we can do is to transfer a direction from one point to another—distant—point *along a specified route.* The result of the transfer will be expected to depend on the route chosen.

Parallel transport is useful for two reasons at least. It is very relevant to the behaviour of an inertial laboratory, and certain delicate experiments depending on it have been proposed. In addition, it provides a clue to the proper understanding of the nature of curvature (for which, see Chapter 7.)

5.2 The Christoffel symbols

At this point we must cast the Euler equations (3.6) in a different form. Since

$$\frac{\partial L}{\partial \dot{x}^\mu} = 2g_{\mu\nu}\dot{x}^\nu \quad \text{and} \quad \frac{\partial L}{\partial x^\mu} = \frac{\partial g_{\alpha\beta}}{\partial x^\mu}\dot{x}^\alpha \dot{x}^\beta,$$

(summation convention) the Euler equations may be written

$$\frac{\mathrm{d}}{\mathrm{d}s}(2g_{\mu\nu}\dot{x}^\nu) - \frac{\partial g_{\alpha\beta}}{\partial x^\mu}\dot{x}^\alpha \dot{x}^\beta = 0.$$

Performing the s-differentiation and raising the suffix μ gives

$$\ddot{x}^\mu + \Gamma^\mu{}_{\alpha\beta}\dot{x}^\alpha \dot{x}^\beta = 0, \tag{5.1}$$

where the **Christoffel symbols** (introduced by E.B. Christoffel in 1869) are defined by

$$\Gamma^{\mu}{}_{\alpha\beta} = \tfrac{1}{2} g^{\mu\nu}\left(\frac{\partial g_{\nu\alpha}}{\partial x^{\beta}} + \frac{\partial g_{\nu\beta}}{\partial x^{\alpha}} - \frac{\partial g_{\alpha\beta}}{\partial x^{\nu}}\right). \tag{5.2}$$

This expression has been adjusted to be *symmetric* in the two lower affixes,

$$\Gamma^{\mu}{}_{\alpha\beta} = \Gamma^{\mu}{}_{\beta\alpha}. \tag{5.3}$$

Any antisymmetric part cancels in (5.1), and is therefore irrelevant.

The expression

$$F^{\mu} = \ddot{x}^{\mu} + \Gamma^{\mu}{}_{\alpha\beta}\dot{x}^{\alpha}\dot{x}^{\beta} \tag{5.4}$$

is zero along a geodesic. It is not zero along other curves, and provides a measure of how far any given curve departs from 'straightness'. We shall not be surprised to find that in General Relativity, **F** yields the **acceleration** of a particle. In fact, **F** is a *contravariant vector*, even though the individual terms are *not* vectors; only the combination has geometrical significance. (This implies of course that the Christoffel symbols themselves are not the elements of any tensor.)

5.3 Calculating the Christoffel symbols

It is often unwise to use the formula to compute the Γs in particular cases. In (1 + 3)D space there are 64 of them (though only 40 are independent on account of the symmetry). In many practical applications, however, the majority are zero. It may then be simpler to go back to the beginning, and to re-obtain the equations (5.1). The nonzero Γs may then be read off as coefficients, while the zero ones never appear at all.

As an example, we shall calculate the (eight) Christoffel symbols for the familiar metric for the surface of a sphere of radius R,

$$\mathrm{d}s^2 = R^2\,\mathrm{d}\theta^2 + R^2 \sin^2\theta\,\mathrm{d}\phi^2.$$

The Lagrangian is

$$L = R^2\dot{\theta}^2 + R^2\sin^2\theta\dot{\phi}^2,$$

from which we obtain the two Lagrangian equations

$$0 = \frac{\mathrm{d}}{\mathrm{d}s}\left(\frac{\partial L}{\partial\dot{\theta}}\right) - \frac{\partial L}{\partial\theta} \equiv \frac{\mathrm{d}}{\mathrm{d}s}(2R^2\dot{\theta}) - 2R^2\sin\theta\cos\theta\dot{\phi}^2,$$

$$0 = \frac{\mathrm{d}}{\mathrm{d}s}\left(\frac{\partial L}{\partial\dot{\phi}}\right) - \frac{\partial L}{\partial\phi} \equiv \frac{\mathrm{d}}{\mathrm{d}s}(2R^2\sin^2\theta\dot{\phi}).$$

On rearrangement, these give

$$F^{\theta} = \ddot{\theta} - \sin\theta\cos\theta\dot{\phi}^2 = 0,$$

and

$$F^{\phi} = \ddot{\phi} + 2\cot\theta\dot{\theta}\dot{\phi} = 0.$$

Inspecting the coefficients with (5.4) in mind shows that three of the symbols are

$$\Gamma^{\theta}{}_{\phi\phi} = -\sin\theta\cos\theta \quad \text{and} \quad \Gamma^{\phi}{}_{\theta\phi} = \Gamma^{\phi}{}_{\phi\theta} = \cot\theta. \tag{5.5}$$

The other five do not appear, and may therefore be taken to be zero. (Note how, on account of symmetry, the contribution of the 'cross term' $2\cot\theta$ is divided equally between *two* of the Christoffel symbols.)

The Christoffel symbols do not depend on R; thus the metric may be scaled up or down at will without affecting the symbols. This is a general feature: the Christoffel symbols are 'scale-free', reflecting the fact that they have much to do with parallelism and angular relationships, and little to do with linear size.

5.4 Parallel transport of a contravariant vector along a smooth curve

Any curve may be specified parametrically by the arc length s (or the interval τ, when appropriate): the coordinates $x^\mu(s)$ of a point on the curve are given as functions of s. At the point A, a vector $\mathbf{v}_A$ is given. At the neighbouring point B the **parallel-transported** $\mathbf{v}_B$ is *defined* by its components

$$v_B^\mu = v_A^\mu - \Gamma^\mu{}_{\alpha\beta}\delta x^\alpha v_A^\beta,$$

using the Christoffel symbols just introduced. Divide by δs and allow $\delta s \to 0$; this gives a differential equation for the continuous parallel transport of $\mathbf{v}$ *along the curve* (Fig. 5.1),

$$\dot{v}^\mu + \dot{x}^\alpha \Gamma^\mu{}_{\alpha\beta} v^\beta = 0. \tag{5.6}$$

It is naturally to be expected that parallel transport from one point to another according to this rule will depend on the path followed, unless the space is flat.

An important special case—partly justifying the whole procedure—is obtained by setting $v^\mu = \dot{x}^\mu$. Then

$$\dot{v}^\mu + \dot{x}^\alpha \Gamma^\mu{}_{\alpha\beta} v^\beta \quad \text{becomes} \quad \ddot{x}^\mu + \Gamma^\mu{}_{\alpha\beta}\dot{x}^\alpha \dot{x}^\beta \equiv F^\mu.$$

Thus the tangent vector of any geodesic (for which of course $F^\mu = 0$) is continuously parallel-transported along the geodesic. Loosely, the geodesic keeps on going in the same direction: it is a *straightest* curve, in addition to being a *shortest*. Thus the promised link with the variational calculation of Chapter 3 is established.

5.5 What happens to a scalar product during parallel transport?

The scalar product of two vectors $\mathbf{a}$ and $\mathbf{b}$ is $g_{\mu\nu}a^\mu b^\nu$. Suppose that $\mathbf{a}$ and $\mathbf{b}$ are parallel-transported together along the curve $\mathbf{x}(s)$, so that

$$\dot{a}^\mu = -\dot{x}^\alpha \Gamma^\mu{}_{\alpha\beta} a^\beta \quad \text{and} \quad \dot{b}^\mu = -\dot{x}^\alpha \Gamma^\mu{}_{\alpha\beta} b^\beta.$$

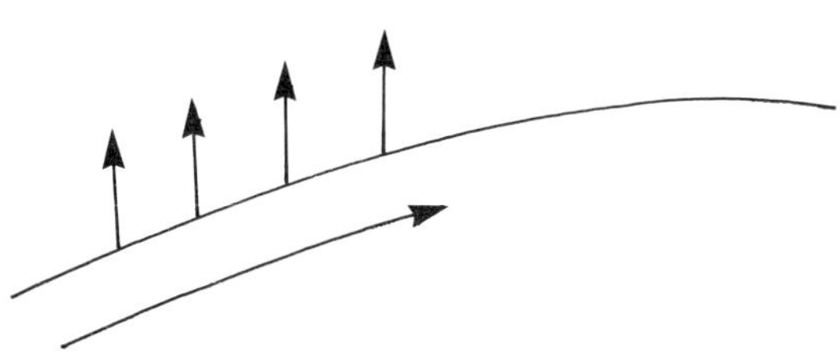

Fig. 5.1

As **x** sweeps along the curve, the metric will change, being a function of position; by the usual rule,

$$\dot{g}_{\mu\nu} = \dot{x}^\sigma \frac{\partial g_{\mu\nu}}{\partial x^\sigma}.$$

Consequently,

$$\frac{\mathrm{d}}{\mathrm{d}s}(g_{\mu\nu}a^\mu b^\nu) = \dot{g}_{\mu\nu}a^\mu b^\nu + g_{\mu\nu}\dot{a}^\mu b^\nu + g_{\mu\nu}a^\mu \dot{b}^\nu$$

$$= a^\mu b^\nu \dot{x}^\sigma \left(\frac{\partial g_{\mu\nu}}{\partial x^\sigma} - \Gamma^\alpha{}_{\mu\sigma} g_{\alpha\nu} - \Gamma^\alpha{}_{\nu\sigma} g_{\mu\alpha} \right)$$

after a slight renaming of the summed affixes. By virtue of the definition of the Γs, the expression in parentheses is identically zero. Thus the scalar product of two simultaneously parallel-transported vectors does not change.

An important special case occurs when **a** and **b** are the same: the *length* of a vector does not change on parallel transport.

5.6 The parallel transport of a covariant vector

The rule for the parallel transport of a covariant vector is now readily obtained. The scalar product $a_\mu b^\mu$ is to be unchanged on parallel transport. That is

$$0 = \frac{\mathrm{d}}{\mathrm{d}s}(a_\mu b^\mu) = \dot{a}_\mu b^\mu + a_\mu \dot{b}^\mu$$

$$= \dot{a}_\mu b^\mu - a_\mu \dot{x}^\sigma \Gamma^\mu{}_{\sigma\alpha} b^\alpha$$

on using (5.6) for $\dot{b}^\mu$. Thus consistency requires (after a slight renaming of affixes)

$$\dot{a}_\mu = \dot{x}^\sigma \Gamma^\alpha{}_{\sigma\mu} a_\alpha. \tag{5.7}$$

These rules are easily generalized to tensors with more than one affix. For example, the result of the parallel transport of the metric tensor would be governed by

$$\dot{g}_{\mu\nu} = \dot{x}^\sigma \Gamma^\alpha{}_{\sigma\mu} g_{\alpha\nu} + \dot{x}^\sigma \Gamma^\alpha{}_{\sigma\nu} g_{\mu\alpha};$$

each and every affix of the tensor attracts just one $\dot{x}\Gamma$-term. We have already seen that this equation is identically satisfied by the *actual* metric tensor: so, if the metric at some point A is parallel-transported along any curve to any other point B, the result is identical to the actual metric at B. Of course, the metric is a very special tensor.

5.7 Parallel transport is really a 'rotation'

Parallel transport preserves lengths of vectors and, because it preserves scalar products, it also preserves angles between vectors. Thus any 'hedgehog' of vectors will rigidly retain its shape and size after parallel transport: in other words, the most that can happen is that it is *rotated* (or *Lorentz-transformed* if the space is (1+3)D) as it moves. (Later, we shall be very interested in the rotation that results from parallel transport round a closed curve, in connection with the idea of curvature.)

Algebraically, a rotation is a linear homogeneous transformation of a certain kind;

this is reflected in the equation for parallel transport, which is linear and homogeneous in **v**. This allows us to introduce a matrix notation which we shall often find useful. Write the contravariant components of **v** as a *column vector* **V**. Then the equation for the parallel transport of **v** along a curve may be rewritten as

$$\dot{\mathbf{V}} = -\dot{x}^\sigma \mathbf{\Gamma}_\sigma \mathbf{V}, \tag{5.8}$$

where the matrices $\mathbf{\Gamma}$ are assembled from the Christoffel symbols by the recipe

$$\Gamma^\alpha{}_{\sigma\beta} = \text{the element in row } \alpha \text{ and column } \beta \text{ of the matrix } \mathbf{\Gamma}_\sigma.$$

Thus the summation convention is here partly replaced by a matrix multiplication. This helps to avoid a proliferation of affixes.

Because, on parallel transport along a curve, the vector $\mathbf{V}(0)$ is subject to a homogeneous linear transformation into the vector $\mathbf{V}(s)$, we may write

$$\mathbf{V}(s) = \mathbf{T}(s)\mathbf{V}(0)$$

where the matrix of the transformation $\mathbf{T}(s)$ must satisfy the differential equation—really a coupled set of such equations—

$$\dot{\mathbf{T}} = -\dot{x}^\sigma \mathbf{\Gamma}_\sigma \mathbf{T} \tag{5.9}$$

and the initial condition

$$\mathbf{T}(0) = \mathbf{I} \quad \text{(the unit matrix).}$$

5.8 An example

To clarify these rather general considerations, think about a particularly simple example: parallel transport in a plane, referred to polar coordinates r and θ.

The metric is

$$\mathrm{d}s^2 = \mathrm{d}r^2 + r^2\,\mathrm{d}\theta^2.$$

Five of the eight Christoffel symbols are zero; the other three are

$$\Gamma^r_{\theta\theta} = -r, \quad \Gamma^\theta_{r\theta} = \Gamma^\theta_{\theta r} = \frac{1}{r}.$$

The eight symbols are to be assembled into two matrices, as follows:

$$\mathbf{\Gamma}_r = \begin{bmatrix} \cdot & \cdot \\ \cdot & 1/r \end{bmatrix} \quad \text{and} \quad \mathbf{\Gamma}_\theta = \begin{bmatrix} \cdot & -r \\ 1/r & \cdot \end{bmatrix}.$$

Let us consider in particular parallel transport round the circle $r = R$ (constant), using length s measured along this path as affine parameter. Evidently

$$\dot{r} = 0, \quad \dot{\theta} = \frac{\varepsilon}{R}$$

in which $\varepsilon = \pm 1$, according to the sense of travel. Thus

$$\dot{x}^\sigma \mathbf{\Gamma}_\sigma \equiv \dot{r}\mathbf{\Gamma}_r + \dot{\theta}\mathbf{\Gamma}_\theta = \varepsilon \begin{bmatrix} \cdot & -1 \\ 1/R^2 & \cdot \end{bmatrix}$$

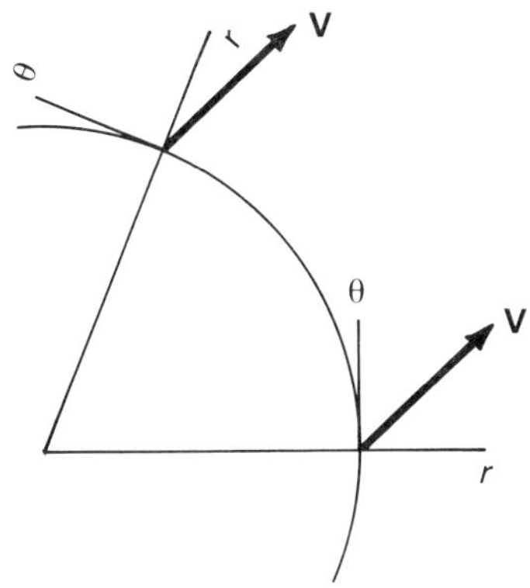

Fig. 5.2

prescribes the relevant right sides of the equations. The 'equations of motion' (5.8) for the contravariant components of a vector under parallel transport along the path are now easily read off:

$$\dot{V}_r = \varepsilon V_\theta, \quad \dot{V}_\theta = -\frac{\varepsilon}{R^2} V_r.$$

The solution is simple harmonic:

$$V_r = A \cos\left(\frac{s}{R} + B\right), \quad V_\theta = -\varepsilon \frac{A}{R} \sin\left(\frac{s}{R} + B\right).$$

The vector **V** itself does not change in magnitude or direction; this is, after all, the meaning of parallel transport in the plane. However, its contravariant components vary, reflecting the changing relation of the vector to the curvilinear coordinates (see Fig. 5.2). As s goes from 0 to 2π, the circle is traversed once, and the components return to their original values.

It should be clear from this example that the Christoffel symbols are primarily concerned with the relationship between the 'real' geometry and the reference system of coordinates. In fact, the values of the symbols at any one point say *nothing* about the geometry itself at that point. It is different for their values considered throughout a finite region: parallel transport from one place to another provides a stepping stone to a proper understanding of curvature.

5.9 Geodesic coordinates at a chosen point

It is always possible to choose coordinates for which all the Γs are zero at any specified point O (though not everywhere else if the space is curved). Finding such coordinates is straightforward. Suppose we already have 'old' coordinates **x**; with no real loss of generality we shall take O to be the origin of these coordinates. We make small (second-order) adjustments in the vicinity of O to obtain 'new' coordinates **y**:

$$y^\mu = x^\mu + \tfrac{1}{2}\Gamma^\mu{}_{\alpha\beta} x^\alpha x^\beta. \tag{5.10}$$

(In this expression it is sufficiently accurate to evaluate the coefficients Γ *at* O *itself*.)

Along any curve through O, we may differentiate twice to obtain

$$\begin{aligned}\ddot{y}^\mu &= \ddot{x}^\mu + \Gamma^\mu{}_{\alpha\beta}\{\tfrac{1}{2}\ddot{x}^\alpha x^\beta + \dot{x}^\alpha\dot{x}^\beta + \tfrac{1}{2}x^\alpha\ddot{x}^\beta\}\\ &= \ddot{x}^\mu + \Gamma^\mu{}_{\alpha\beta}\dot{x}^\alpha\dot{x}^\beta, \text{ when evaluated at the point O.}\end{aligned}$$

Comparing the right side with the general equation for a geodesic shows that for any geodesic through O,

$$\ddot{y}^\mu = 0 \text{ at the point O.}$$

In the new **geodesic coordinates y**, at O, the geodesic equations contain no Christoffel terms; thus all the Γs have been transformed to zero.

It follows, by the way, that the Christoffel symbols are *not* the elements of a tensor: if they were, then the fact that they are zero in one system of coordinates would imply that they are zero in all, in conflict with what we have just done.

5.10 Geodesic coordinates along a geodesic

We can do even better, in a way that has important implications for the Principle of Equivalence. It is possible to arrange for the Γs to be zero at all points along any chosen curve; the most important instance occurs when the curve is a *geodesic*. The recipe considers a narrow tube enclosing the geodesic, and fills it with a suitable coordinate system. Here it is, without proof:

(i) The distance s along the geodesic is to be one coordinate.
(ii) The orthogonal cross-section of the tube at $s=0$ (say) is provided with an approximately Cartesian set of coordinates z^1, z^2, z^3, with origin on the geodesic itself.
(iii) This coordinate mesh at $s=0$ is extended throughout the length of the tube by parallel transport along the geodesic.

This is already enough to provide coordinates for the entire tube, and to ensure that a large proportion of the Γs is zero. Making the remaining Γs zero requires a second-order 'straightening' of the coordinates: this final step is, in the style of (5.10),

(iv) Make second-order local adjustments to *new* coordinates S, Z^1, Z^2, Z^3 according to

$$S = s + \tfrac{1}{2}\sum_{k=1}^{3}\sum_{l=1}^{3}\Gamma^0{}_{kl}z^k z^l,$$

$$Z^m = z^m + \tfrac{1}{2}\sum_{k=1}^{3}\sum_{l=1}^{3}\Gamma^m{}_{kl}z^k z^l.$$

This recipe provides a system of coordinates, valid throughout a tube enclosing a specified geodesic; throughout the tube *all the Γs are small, and actually precisely zero on the geodesic itself.*

5.11 The Principle of Equivalence

The last result implies that a narrow tube of spacetime enclosing a geodesic worldline is more or less the same as a narrow tube surrounding a *straight* worldline in flat

Minkowski spacetime—'more or less' since the correspondence is not quite perfect if the actual spacetime is curved. Thus, from an experimental point of view, an inertial observer—whose worldline is a geodesic—will expect to find that, for him, all the laws of Special Relativity apply: the worldlines of his immediate surroundings will fill a tube which appears to him *to be almost exactly the same as a straight tube excised from flat spacetime.*

In detail, the recipe of the last section has the following relevance to an inertial observer. Theoretically, the worldline of such an observer is to be a geodesic, and since he is not rotating, any coordinate mesh fixed inside his laboratory is to be parallel-transported along this geodesic; the relevance of step (iii) is now clear. The adjustment to s in step (iv) has to do with the synchronization (as far as possible) of the clocks in the different reaches of the laboratory, while the adjustment to the zs straightens out any remaining curvature in the coordinate grid in the laboratory (the coordinates are now as *Cartesian* as they can be). The outcome is that with respect to such coordinates and such timekeeping, free objects in the laboratory are seen to move without acceleration: *the laboratory is inertial in a real experimental sense.*

It is impossible to overstress the importance of this result, since it shows that we are allowed to have inertial observers within the theoretical framework of General Relativity. If this were not so, theory and experience would be in conflict at a fundamental level.

To summarize: An inertial laboratory is a freely falling non-rotating laboratory. In view of the nature of the recipe, we assert for an inertial laboratory that

(i) the worldline of the laboratory centre is a timelike geodesic,
(ii) the laboratory clock measures interval along this geodesic, and
(iii) fixed Cartesian axes in the laboratory are parallel-transported along the geodesic.

Then we have the Principle of Equivalence in an explicit form:

In any inertial laboratory, all of Special Relativity applies.

The Principle makes possible an immediate generalization of any physical phenomenon in Special Relativity to a corresponding phenomenon in General Relativity. How this is done will be discussed later. There is the expected small-print proviso, however. In a large enough laboratory the curvature of spacetime may become noticeable, manifesting itself as an extra tidal force. It is conceivable that the translation of the Special version of a phenomenon to its General version will require the tidal force to be taken into account. We have already met an example: a freely falling charged body may radiate (section 1.13).

5.12 The Christoffel symbols for the Schwarzschild metric

Once again we look at the equations for a geodesic in the Schwarzschild metric; only this time the purpose is to determine the Christoffel symbols.

The Lagrangian is

$$L = c^2(1 - a/r)\dot{t}^2 - \frac{\dot{r}^2}{1 - a/r} - r^2\dot{\theta}^2 - r^2 \sin^2 \theta\dot{\phi}^2. \tag{5.11}$$

The 't-equation' is

$$0=\frac{\mathrm{d}}{\mathrm{d}\tau}\left(\frac{\partial L}{\partial \dot{t}}\right)-\frac{\partial L}{\partial t}=\frac{\mathrm{d}}{\mathrm{d}\tau}[2c^2(1-a/r)\dot{t}]$$
$$=2c^2(1-a/r)\ddot{t}+2c^2(a/r^2)\dot{r}\dot{t}.$$

Dividing through by the coefficient of the second derivative gives the left side in the required style of (5.4),

$$F^t=\ddot{t}+\frac{a}{(1-a/r)r^2}\dot{r}\dot{t}. \tag{5.12a}$$

We may immediately read off *two* nonzero Christoffel symbols,

$$\Gamma^t{}_{tr}=\Gamma^t{}_{rt}=\frac{a}{2(1-a/r)r^2}.$$

The remaining three left sides are similarly obtained in the required style; they are

$$F^r=\ddot{r}+\frac{ac^2}{2r^2}(1-a/r)\dot{t}^2-\frac{a}{2r^2(1-a/r)}\dot{r}^2-r(1-a/r)\dot{\theta}^2-r(1-a/r)\sin^2\theta\dot{\phi}^2, \tag{5.12b}$$

$$F^\theta=\ddot{\theta}+\frac{2}{r}\dot{r}\dot{\theta}-\sin\theta\cos\theta\dot{\phi}^2, \tag{5.12c}$$

$$F^\phi=\ddot{\phi}+\frac{2}{r}\dot{r}\dot{\phi}+2\cot\theta\dot{\theta}\dot{\phi}. \tag{5.12d}$$

The complete list of thirteen nonzero Christoffel symbols may now be read off from

$$\Gamma_t=\begin{bmatrix} \cdot & a/2r^2f & \cdot & \cdot \\ ac^2f/2r^2 & \cdot & \cdot & \cdot \\ \cdot & \cdot & \cdot & \cdot \\ \cdot & \cdot & \cdot & \cdot \end{bmatrix} \qquad \Gamma_r=\begin{bmatrix} a/2r^2f & \cdot & \cdot & \cdot \\ \cdot & -a/2r^2f & \cdot & \cdot \\ \cdot & \cdot & 1/r & \cdot \\ \cdot & \cdot & \cdot & 1/r \end{bmatrix}$$

$$\Gamma_\theta=\begin{bmatrix} \cdot & \cdot & \cdot & \cdot \\ \cdot & \cdot & -rf & \cdot \\ \cdot & 1/r & \cdot & \cdot \\ \cdot & \cdot & \cdot & \cot\theta \end{bmatrix} \qquad \Gamma_\phi=\begin{bmatrix} \cdot & \cdot & \cdot & \cdot \\ \cdot & \cdot & \cdot & -rf\sin^2\theta \\ \cdot & \cdot & \cdot & -\sin\theta\cos\theta \\ \cdot & 1/r & \cot\theta & \end{bmatrix}$$

$$f\equiv 1-a/r$$

Fig. 5.3

the nine coefficients in the four expressions. With $f \equiv 1 - a/r$ for brevity, they are

$$\Gamma^t{}_{tr} = \Gamma^t{}_{rt} = \frac{a}{2r^2 f}$$

$$\Gamma^r{}_{tt} = \frac{ac^2 f}{2r^2} \qquad \Gamma^r{}_{rr} = -\frac{a}{2r^2 f} \qquad \Gamma^r{}_{\theta\theta} = -rf \qquad \Gamma^r{}_{\phi\phi} = -rf\sin^2\theta$$

$$\Gamma^\theta{}_{r\theta} = \Gamma^\theta{}_{\theta r} = \frac{1}{r} \qquad \Gamma^\theta{}_{\phi\phi} = -\sin\theta\cos\theta$$

$$\Gamma^\phi{}_{r\phi} = \Gamma^\phi{}_{\phi r} = \frac{1}{r} \qquad \Gamma^\phi{}_{\theta\phi} = \Gamma^\phi{}_{\phi\theta} = \cot\theta. \tag{5.13}$$

Their assembly into the four matrices Γ_t, Γ_r, Γ_θ, Γ_ϕ is displayed in Fig. 5.3.

5.13 The forces acting on an Earth-bound laboratory

A laboratory is fixed at colatitude θ_0, being carried along by the rotation of the Earth. With laboratory proper time τ as parameter, its worldline in Schwarzschild coordinates is

$$t = \alpha\tau, \quad r = R \text{ (const)}, \quad \theta = \theta_0 \text{ (const)}, \quad \phi = \omega\tau.$$

Evidently the contravariant velocity four-vector is

$$\dot{t} = \alpha, \quad \dot{r} = 0, \quad \dot{\theta} = 0, \quad \text{and } \dot{\phi} = \omega.$$

All second derivatives are zero. In order that the four-velocity should be a 'unit' vector, we require

$$c^2 = \dot{x}^\mu \dot{x}^\nu g_{\mu\nu} = c^2 f\alpha^2 - r^2\omega^2 \sin^2\theta_0.$$

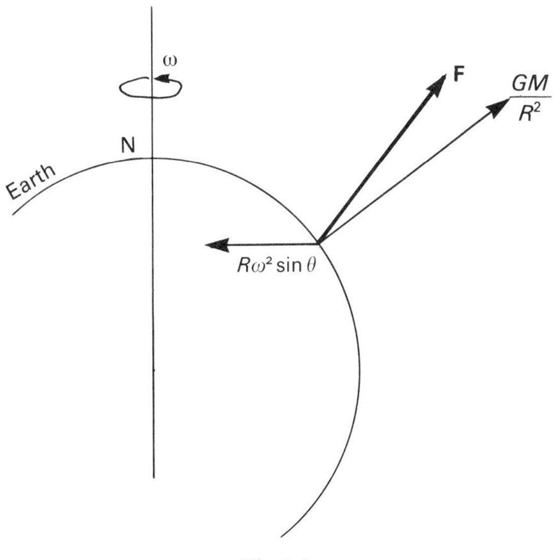

Fig. 5.4

We shall calculate the acceleration vector

$$F^{\mu} = \ddot{x}^{\mu} + \Gamma^{\mu}{}_{\alpha\beta}\dot{x}^{\alpha}\dot{x}^{\beta}.$$

In fact, the components are the four expressions (5.12a)–(5.12d)—no longer zero, of course, since the worldline under consideration is not a geodesic. All the second derivatives are zero, and only three of the remaining terms survive; in fact,

$$\begin{aligned} F^t &= 0 \\ F^r &= \frac{ac^2 f}{2R^2}\alpha^2 - Rf\sin^2\theta\omega^2 \\ F^{\theta} &= -\sin\theta\cos\theta\omega^2 \\ F^{\phi} &= 0. \end{aligned}$$

We are dealing with relatively slow motion, and α may be replaced by 1; f is very nearly 1 anyway. Additionally, ac^2 is to be interpreted as $2GM$, where M is the mass of the Earth (see section 4.8) (Fig. 5.4). After these replacements,

$$\begin{aligned} F^r &= \frac{GM}{R^2} - R\omega^2\sin^2\theta \\ RF^{\theta} &= -R\omega^2\sin\theta\cos\theta \end{aligned}$$

are the components of the force (per unit mass) needed to keep the laboratory where it is; this force is the reaction of the ground on the foundations, needed to counterbalance the kinematic centrifugal and gravity forces. Note how the two apparently different kinds of force arise together in the development, showing that in the framework of General Relativity the Eötvös experiment is guaranteed to give a null result.

5.14 The geodesic effect

Laboratories in free fall used to be hard to come by, but they have become commonplace in the age of artificial satellites. New experiments to test the consequences of parallel transport are thus now feasible. What happens when an inertial laboratory is parallel-transported round a closed planetary orbit is of particular interest.

Fig. 5.5 shows qualitatively what may be expected to happen. A satellite is placed in a low equatorial circular earth orbit: its worldline in spacetime is a helical geodesic. (Fig. 5.5 depicts only events which occur on the equatorial plane $\theta = \pi/2$, and is therefore $(1+2)$D only.) If the satellite is inertial—and therefore non-rotating—then axes fixed in the satellite will be parallel-transported along the helix: there is clearly no reason to expect that after one orbit these axes will still have the same orientation as before in relation to the Schwarzschild coordinate mesh. They may have *rotated*. Experimentally, therefore, an inertial guidance system—based on accelerometers—mounted in the satellite may show a small drift when compared with an optical guidance system based on the distant 'fixed' stars.

The worldline for a circular orbit may be described parametrically by

$$t = \alpha\tau, \quad r = \text{const}, \quad \theta = \pi/2, \quad \phi = \omega\tau,$$

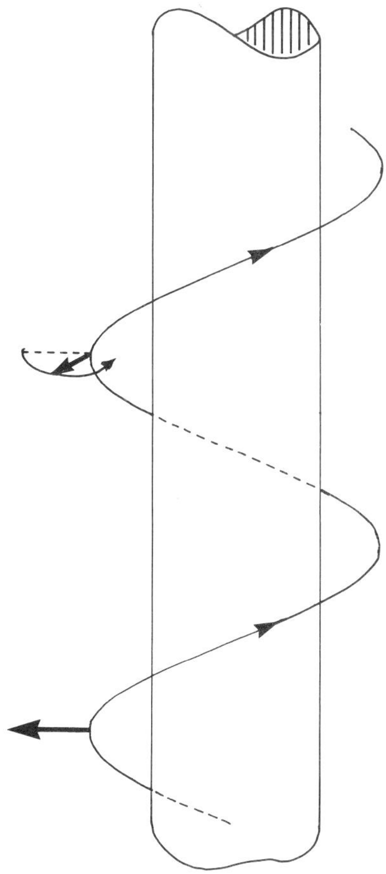

Fig. 5.5

where τ is the proper time by the satellite clock, and α, ω are constants. The four-velocity vector is

$$\dot{x}^\mu = (\dot{t}, \dot{r}, \dot{\theta}, \dot{\phi}) = (\alpha, 0, 0, \omega)$$

and the contribution of the Γ-symbols (from Fig. 5.3) reduces to

$$\dot{x}^\sigma \mathbf{\Gamma}_\sigma = \alpha \mathbf{\Gamma}_t + \omega \mathbf{\Gamma}_\phi = \begin{bmatrix} \cdot & \alpha a/2fr^2 & \cdot & \cdot \\ \alpha ac^2 f/2r^2 & \cdot & \cdot & -rf\omega \\ \cdot & \cdot & \cdot & \cdot \\ \cdot & \omega/r & \cdot & \cdot \end{bmatrix}.$$

(Note that $f = 1 - a/r$, and θ has been set to $\pi/2$ throughout.) This matrix is clearly *constant* for a circular orbit—though different for different radii—and the equations of motion are therefore easily solved. They are

$$\frac{\mathrm{d}}{\mathrm{d}\tau}\begin{bmatrix} \lambda^t \\ \lambda^r \\ \lambda^\theta \\ \lambda^\phi \end{bmatrix} = -\begin{bmatrix} \cdot & \alpha a/2fr^2 & \cdot & \cdot \\ \alpha ac^2 f/2r^2 & \cdot & \cdot & -rf\omega \\ \cdot & \cdot & \cdot & \cdot \\ \cdot & \omega/r & \cdot & \cdot \end{bmatrix}\begin{bmatrix} \lambda^t \\ \lambda^r \\ \lambda^\theta \\ \lambda^\phi \end{bmatrix},$$

where $\boldsymbol{\lambda}$ is any vector parallel-transported along the worldline of the satellite. These equations form a linear, simultaneous set, homogeneous in the components of $\boldsymbol{\lambda}$ and $\dot{\boldsymbol{\lambda}}$, all coefficients being constant. The general solution of such equations may always be expressed as a linear combination of **normal mode** solutions $\boldsymbol{\lambda} = \boldsymbol{\lambda}_0 \exp \mathrm{i}\Omega\tau$, where for each normal mode, $\boldsymbol{\lambda}_0$ is a certain constant vector, and Ω is a certain angular frequency.

The tangent vector $\dot{\mathbf{x}} = (\alpha, 0, 0, \omega)$ to the worldline must satisfy these equations, since the worldline is a geodesic. Substitution leads directly to just one requirement

$$\frac{\alpha a c^2 f}{2r^2}\alpha - rf\omega \cdot \omega = 0,$$

that is,

$$r^3\omega^2 = \tfrac{1}{2}ac^2\alpha^2,$$

relating the radius of the orbit to its period (one of Kepler's conditions for planetary motion). This in fact gives one of the normal modes, with $\Omega = 0$.

Substituting $\boldsymbol{\lambda} = \boldsymbol{\lambda}_0 \exp \mathrm{i}\Omega\tau$ in the equations reduces the problem to the algebraic solution of four linear simultaneous equations, homogeneous in the four components of $\boldsymbol{\lambda}_0$. It is a familiar fact that unless a certain condition is satisfied, the only solution is the trivial one $\boldsymbol{\lambda}_0 = 0$: this condition is that the determinant of the equations should vanish,

$$\begin{vmatrix} \mathrm{i}\Omega & \alpha a/2fr^2 & \cdot & \cdot \\ \alpha a c^2 f/2r^2 & \mathrm{i}\Omega & \cdot & -rf\omega \\ \cdot & \cdot & \mathrm{i}\Omega & \cdot \\ \cdot & \omega/r & \cdot & \mathrm{i}\Omega \end{vmatrix} = 0.$$

This is

$$\Omega^4 + \Omega^2\omega^2\left(\frac{\alpha^2 a^2 c^2}{4r^4} - f\omega^2\right) = 0.$$

Using the Kepler condition reduces this to

$$\Omega^4 - \Omega^2\omega^2(1 - 3a/2r) = 0.$$

There are two zero roots: one is inevitable, declaring that $\dot{\mathbf{x}}$ must satisfy the equations, while the other is obvious, being associated with the constant solution $\boldsymbol{\lambda} = (0, 0, \text{constant}, 0)$. ('Obvious' on account of equatorial symmetry.) The other two roots are given by $\pm\Omega$, with

$$\Omega = \omega\sqrt{(1 - 3a/2r)}.$$

A *real* solution to the original equations may now be obtained in the form

$$\boldsymbol{\lambda} = \boldsymbol{\lambda}_0 \exp \mathrm{i}\Omega\tau + \boldsymbol{\lambda}_0^* \exp - \mathrm{i}\Omega\tau$$

and it is not difficult to verify that this is very nearly a *rotation* of the vector $\boldsymbol{\lambda}$ with respect to the $tr\theta\phi$-system of coordinates with angular velocity Ω. Of course, as the satellite itself traverses its orbit, the $tr\theta\phi$-system itself rotates with angular velocity ω, slightly larger and in the opposite sense. Thus $\boldsymbol{\lambda}$ experiences a small net rotation with respect to distant parts of the Universe; the angular velocity of this rotation is

$$\omega - \Omega = \frac{3a}{4r}\omega + O(a^2)$$

in the same sense as the description of the orbit. This gives the 'drift' of an inertial guidance system based on accelerometers or their equivalent, as compared with a guidance system based on sightings of distant astronomical objects. The effect is cumulative, and for a low Earth orbit amounts to about 8 seconds of arc per year, probably just observable with present technology. There are plans to launch a suitable satellite.

The result is usually referred to as the **geodesic effect**, though this is something of a misnomer, since it may be expected for any worldline, not just a geodesic. The effect was predicted and calculated in 1921 by A.D. Fokker, not of course for a satellite in Earth gravity, but for the Earth itself in its orbit round the Sun; in that case the drift is 0.02 seconds of arc per year. At that time, of course, it was felt that there was no hope whatever of any verification.

5—Notes and problems

1. Along a *general* curve with affine parameter s, the tangent vector is of course *not* parallel-transported, and therefore $\mathbf{F} \neq 0$.

   However, the length of the tangent vector is constant. Hence show that, along *any* curve,

$$0 = \frac{\mathrm{d}}{\mathrm{d}\sigma}(\dot{x}^\mu \dot{x}^\nu g_{\mu\nu}) \equiv 2\dot{x}^\mu F_\mu.$$

   Thus the curvature vector (or acceleration vector, according to context) is orthogonal to the tangent; it is the **normal** to the curve.

   [In doing the differentiation with respect to s, second derivatives of $\mathbf{x}$ appear; these are to be eliminated with the help of (5.4). There results

$$\frac{\mathrm{d}}{\mathrm{d}s}(\dot{x}^\mu \dot{x}^\nu g_{\mu\nu}) = 2F_\mu \dot{x}^\mu + \dot{x}^\alpha \dot{x}^\beta \dot{x}^\gamma \left(-2\Gamma^\mu{}_{\alpha\beta} g_{\mu\gamma} + \frac{\partial}{\partial x^\gamma} g_{\alpha\beta} \right).$$

   The second part of the right side is zero, on account of the definition (5.2) of the Γs and of the symmetry between the dummy indices $\alpha\beta\gamma$.]

2. Evaluate the Christoffel symbols for each of the metrics in Problem 1 of Chapter 3. (The answers for two cases are

   (iii): $\Gamma^u{}_{uu} = \frac{1}{2}$ $\Gamma^u{}_{uv} = \Gamma^u{}_{vu} = -\frac{1}{2}$ $\Gamma^u{}_{vv} = -\frac{1}{2}$

   $\Gamma^v{}_{uu} = \frac{1}{2}$ $\Gamma^v{}_{uv} = \Gamma^v{}_{vu} = \frac{1}{2}$ $\Gamma^v{}_{vv} = -\frac{1}{2}$

   (vi): $\Gamma^w{}_{\phi\phi} = \mathrm{e}^{-2w}$ $\Gamma^\phi{}_{w\phi} = \Gamma^\phi{}_{\phi w} = -1$; remainder zero.)

3. (i) In Cartesian coordinates x, y in the plane, as s goes from 0 to $2\pi/R$, the point

$$x = R\cos s/R, \quad y = R\sin s/R$$

   traverses a circle of radius R centred on the origin. Since all the Christoffel symbols are zero in this case,

$$F^x = \ddot{x} = -\frac{1}{R}\cos\frac{s}{R}, \quad F^y = \ddot{y} = -\frac{1}{R}\sin\frac{s}{R}.$$

Thus **F** is an inwardly directed vector with magnitude $1/R$, R being the radius of curvature of the circle.

(ii) In the corresponding polar coordinates r, θ, as s goes from 0 to $2\pi/R$,

$$r = R, \quad \theta = s/R$$

traverses the same circle. In this case, the nonzero Christoffel symbols are

$$\Gamma^r{}_{\theta\theta} = -r, \quad \Gamma^\theta{}_{r\theta} = \Gamma^\theta{}_{\theta r} = \frac{1}{r},$$

and the second derivatives are zero, as is the first derivative of $r = R$. All that now survives is

$$F^r = \Gamma^r{}_{\theta\theta}\dot{\theta}\dot{\theta} = -R\frac{1}{R^2}, \quad F^\theta = 0.$$

Verify that **F** is an inwardly directed vector with magnitude $1/R$, R being the radius of curvature of the circle.

The algebra in the two cases looks very different, but the final geometrical significance is the same. It is not helpful to try to interpret the individual terms in the calculation of the components of **F**; only the complete expression is of any interest.

4. On the surface of a sphere of radius R, with the usual coordinates θ and ϕ, the parallel of colatitude θ_0 is traversed by

$$\theta = \theta_0, \quad \phi = \frac{s}{R\sin\theta_0}$$

as s varies. Show that

$$F^\theta = -\frac{1}{R^2}\cot\theta, \quad F^\phi = 0,$$

and that **F** is a vector of magnitude $R^{-1}\cot\theta$ directed towards the pole. Section 5.3 contains relevant information.

(Note carefully that this means that the circle has radius of curvature $R\tan\theta$ when considered as embedded in the surface of the sphere. This must be contrasted with the more 'obvious' radius $R\sin\theta$. In particular, the Equator has the 'obvious' radius R, but considered as a geodesic on the sphere it is utterly straight. This is of course an instance of the necessary distinction between *intrinsic* and *extrinsic*.)

5. The vector **V** is parallel-transported once round the parallel of latitude $\theta = \theta_0$ on the surface of a sphere of radius R; as in the previous problem,

$$\theta = \theta_0, \quad \phi = \frac{s}{R\sin\theta_0}$$

and s goes from 0 to $2\pi R \sin\theta_0$. On the lines of section 5.8, show that the change in the components of $\mathbf{V}$ is governed by

$$\begin{bmatrix} \dot{V}_\theta \\ \dot{V}_\phi \end{bmatrix} = -\frac{1}{R\sin\theta_0}\begin{bmatrix} \cdot & -\sin\theta_0\cos\theta_0 \\ \cot\theta_0 & \cdot \end{bmatrix}\begin{bmatrix} V_\theta \\ V_\phi \end{bmatrix}.$$

(The coefficients are constant; the changes are therefore simple harmonic.) Show that after the operation, $\mathbf{V}$ will have changed orientation (unless the transport is round the Equator), the angle of change being $2\pi\cos\theta_0$.

The corresponding value for a flat surface is 2π. (Why not zero?) The discrepancy $2\pi(1-\cos\theta_0)$ is the **spherical excess** associated with the closed curve, and is a measure of the total curvature enclosed. In fact, the area enclosed by the parallel of latitude is exactly $2\pi R^2(1-\cos\theta_0)$, and we take as a definition (in 2D *only*)

$$\text{mean intrinsic curvature} = \frac{\text{spherical excess}}{\text{area}} = \frac{1}{R^2}.$$

The *intrinsic curvature at a point* will be obtained in the limit of a vanishing area element enclosing the point; in the present case this is $1/R^2$. Curvature in this sense always has dimensions of $(\text{area})^{-1}$.

Later we shall need to extend these ideas to more than two dimensions, where things are not so simple.

6. On the surface of a sphere, verify that θ and ϕ are geodesic coordinates along the Equator. (It is sufficient to note that all the Christoffel symbols are zero on the Equator.)

   The coordinates α and β (Chapter 3, Problem 2) are geodesic along the meridians $\phi = \pm\pi/2$.

7. An Earth-bound laboratory is travelling due south on a smooth, level railway track with constant speed v, so that

$$t = \gamma\tau, \quad r = R \text{ (constant)}, \quad \theta = v\tau/R, \quad \phi = \omega\tau.$$

Show that calculating F^μ now produces two further terms, which compensate for an extra centrifugal force (proportional to v^2) and the **Coriolis force** (proportional to v).

6

The Riemann tensor

Everything should be made as simple as possible, but not simpler—Albert Einstein.

The geometrical object which describes intrinsic curvature is the *Riemann tensor*. In this chapter the origins and the properties of this tensor will be derived. The precise meaning of intrinsic curvature, and the way in which the Riemann tensor relates to it, will be dealt with in Chapter 7.

6.1 Covariant derivatives

A **scalar field** $\phi(\mathrm{P})$ is a *scalar function of position*, that is, a scalar defined at each and every point P of some space. If P and Q are neighbouring points, the change $\phi(\mathrm{Q})-\phi(\mathrm{P})$ is also a scalar. In a chosen coordinate system, the displacement PQ will have the contravariant components $\mathrm{d}x^\mu$, and by the usual rules for partial differentiation

$$\phi(\mathrm{Q})-\phi(\mathrm{P}) = \mathrm{d}x^\mu \frac{\partial\phi}{\partial x^\mu} + O(\mathrm{d}x^2)$$

$$\equiv \mathrm{d}x^\mu \partial_\mu \phi + O(\mathrm{d}x^2)$$

(summed, of course). Thus, on account of the way it associates with a contravariant vector, $\partial_\mu \phi$ is a covariant vector, the **gradient** of ϕ at P. [Note the abbreviated notation. We shall need to use partial derivatives extensively from now on, and it is helpful to use the more compact symbolism

$$\partial_\mu \equiv \frac{\partial}{\partial x^\mu} \tag{6.1}$$

throughout. In other texts, a *comma notation* is often used; for example

$$\phi_{,\mu} \equiv \partial_\mu \phi.$$

We shall not use it in this book.]

Now the idea behind a derivative is that quantities at two closely neighbouring points are being compared. For example, for a scalar field ϕ and a small vector $\varepsilon\mathbf{h}$, a typical definition of a derivative is

$$\lim_{\varepsilon\to 0}\frac{\phi(\mathbf{x}+\varepsilon\mathbf{h})-\phi(\mathbf{x})}{\varepsilon}=h^{\nu}\partial_{\nu}\phi.$$

This is perfectly in order for a scalar field, but the corresponding problem for a vector field in a curved space hits the snag that vectors at different points (even at neighbouring points) cannot be directly compared, since there may be no universal parallelism to refer their directions to. The solution is to compare vectors at the *same* place: the actual vector $\mathbf{v}(\mathbf{x}+\varepsilon\mathbf{h})$ is compared with the result of parallel-transporting the vector $\mathbf{v}(\mathbf{x})$ along the small displacement $\varepsilon\mathbf{h}$ (Fig. 6.1). Thus we write

$$\lim_{\varepsilon\to 0}\frac{v^{\alpha}(\mathbf{x}+\varepsilon\mathbf{h})-[v^{\alpha}(\mathbf{x})-\varepsilon\mathbf{h}^{\sigma}\Gamma^{\alpha}{}_{\sigma\rho}v^{\rho}(\mathbf{x})]}{\varepsilon}=h^{\sigma}\{\partial_{\sigma}v^{\alpha}+\Gamma^{\alpha}{}_{\sigma\rho}v^{\rho}\}.$$

The right side is a vector for all vectors $\mathbf{h}$, and therefore

$$v^{\alpha}{}_{;\sigma}\equiv\partial_{\sigma}v^{\alpha}+\Gamma^{\alpha}{}_{\sigma\rho}v^{\rho}\tag{6.2}$$

is a tensor, the **covariant derivative** of the contravariant vector $\mathbf{v}$. [The *semicolon notation* is universally used; any suffix which follows a semicolon is taken to imply the operation of covariant differentiation.] A similar discussion may be used for the covariant derivative of a covariant vector field $\mathbf{u}$; we find

$$u_{\alpha;\sigma}\equiv\partial_{\sigma}u_{\alpha}-\Gamma^{\rho}{}_{\sigma\alpha}u_{\rho}.\tag{6.3}$$

This sets the scene for the general case; each contravariant (covariant) affix attracts one extra positive (negative) Γ-term in the covariant derivative, exactly as in the case of parallel transport (where, after all, the whole idea comes from; section 5.4). An example should make this clear:

$$A^{\mu}_{\nu;\sigma}\equiv\partial_{\sigma}A^{\mu}_{\nu}+\Gamma^{\mu}{}_{\sigma\alpha}A^{\alpha}_{\nu}-\Gamma^{\alpha}{}_{\sigma\nu}A^{\mu}_{\alpha}.$$

It is worth mentioning that the notation is deceptively compact. By the summation convention, what we have here (in 4D) is a set of 64 equations, each with nine terms on the right side. Typically, it is much easier in the context of General Relativity to talk in general terms than to carry out the explicit algebra for a particular case. Sometimes the algebra is of astronomical proportions, and is performed these days with the help of a suitably programmed computer.

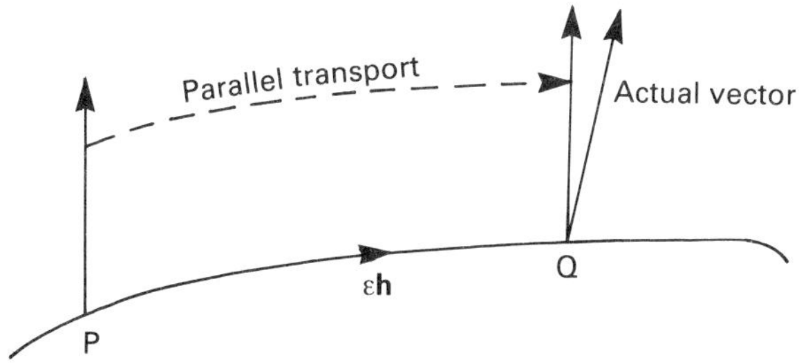

Fig. 6.1

6.2 Covariant differentiation of a product

It is straightforward to show that the rule for the ordinary derivative of a product carries over unchanged for the covariant derivative. For example,

$$(v^\mu u_\nu)_{;\sigma} \equiv v^\mu{}_{;\sigma} u_\nu + v^\mu u_{\nu;\sigma}.$$

An important application of this rule relies on the fact that the covariant derivative of the metric

$$g_{\mu\nu;\sigma} \equiv \partial_\sigma g_{\mu\nu} - \Gamma^\alpha{}_{\sigma\mu} g_{\alpha\nu} - \Gamma^\alpha{}_{\sigma\nu} g_{\mu\alpha} \tag{6.4}$$

is always *identically zero* on account of the original definition of the Γs. The consequence is that the operation of raising or lowering an affix *invariably commutes with covariant differentation.* For example,

$$a_{\mu;\sigma} = g_{\mu\nu} a^\nu{}_{;\sigma}.$$

6.3 The Riemann tensor

The second covariant derivative of a scalar is straightforward to write down (note that it is usual to omit extra semicolons when the effect is unambiguous):

$$\phi_{;\mu\nu} \equiv (\phi_{;\mu})_{;\nu} = \partial_\nu \phi_{;\mu} - \Gamma^\alpha{}_{\nu\mu} \phi_{;\alpha} = \partial_\mu \partial_\nu \phi - \Gamma^\alpha{}_{\nu\mu} \partial_\alpha \phi, \tag{6.5}$$

The Γ-term is present on account of the first derivative being a vector. Each term on the right is unchanged against interchange of μ and ν, for different reasons; hence the *second covariant derivative of a scalar is symmetric*:

$$\phi_{;\mu\nu} \equiv \phi_{;\nu\mu}. \tag{6.6}$$

The second covariant derivative of a vector is more tedious:

$$\begin{aligned} a_{\mu;\rho\sigma} &\equiv (a_{\mu;\rho})_{;\sigma} \\ &= \partial_\sigma(\partial_\rho a_\mu - \Gamma^\alpha{}_{\rho\mu} a_\alpha) - \Gamma^\beta{}_{\sigma\rho}(\partial_\beta a_\mu - \Gamma^\alpha{}_{\beta\mu} a_\alpha) - \Gamma^\beta{}_{\sigma\mu}(\partial_\rho a_\beta - \Gamma^\alpha{}_{\rho\beta} a_\alpha) \\ &= (\text{terms symmetric in } \rho \text{ and } \sigma) - a_\alpha \partial_\sigma \Gamma^\alpha{}_{\rho\mu} + \Gamma^\beta{}_{\sigma\mu} \Gamma^\alpha{}_{\rho\beta} a_\alpha. \end{aligned}$$

We see that in this case the second covariant derivative need not be symmetric: in fact, its antisymmetric part is given by

$$a_{\mu;\rho\sigma} - a_{\mu;\sigma\rho} = a_\alpha R^\alpha{}_{\mu\rho\sigma}, \tag{6.7}$$

where the rogue terms involve the **Riemann tensor**

$$R^\alpha{}_{\mu\rho\sigma} \equiv \partial_\rho \Gamma^\alpha{}_{\sigma\mu} - \partial_\sigma \Gamma^\alpha{}_{\rho\mu} + \Gamma^\alpha{}_{\rho\beta} \Gamma^\beta{}_{\sigma\mu} - \Gamma^\alpha{}_{\sigma\beta} \Gamma^\beta{}_{\rho\mu}. \tag{6.8}$$

In 4D, this defines the 256 elements of the Riemann tensor, with ten terms on each right side, by the summation convention. We shall need to have an organized technique for their evaluation.

The Riemann tensor calls for careful attention. In a flat space, the second covariant derivative must be symmetric. (To see this, use Cartesian coordinates, for which the covariant derivative reduces to the ordinary derivative. Symmetry is an invariant property; it persists when the coordinate system is changed.) Thus if ever the Riemann tensor is nonzero we shall know that we must be dealing with a space which is not flat, and some at least of the information about the curvature is embodied in the nonzero Riemann tensor. In fact, it will turn out that the Riemann tensor tells us *everything essential* about intrinsic curvature, and it is therefore of fundamental importance. Any further curvature arising from an imagined embedding in a space of higher dimension is ignored by the Riemann tensor, which therefore takes us to the heart of the matter.

6.4 The symmetries of the Riemann tensor

It often happens that the elements of a tensor exhibit a property (a **symmetry**) against a permutation of suffixes. The elements of the metric tensor, for example, do not change when the covariant suffixes are interchanged:

$$g_{\mu\nu} = g_{\nu\mu}:$$

the metric tensor is a *symmetric* tensor. It is an important fact that *a symmetry of this kind always persists under a transformation of coordinates*, and that it is therefore a geometrical property of the tensor itself. We may sensibly talk about the symmetries of a tensor, without reference to any particular system of coordinates.

The Riemann tensor always satisfies a number of important symmetries, which spring from its close connection with covariant differentiation. The first of these is

$$\textbf{antisymmetry in the last pair of suffixes:}\quad R^{\alpha}{}_{\mu\rho\sigma} \equiv -R^{\alpha}{}_{\mu\sigma\rho} \tag{6.9}$$

and follows directly from the definition.

The next symmetry may be obtained by a fairly simple manipulation of the original defining relation. We use the *third* covariant derivative of a scalar ϕ, permuting the indices in a variety of ways. Since the second covariant derivative of a scalar is symmetric,

$$-\phi_{;\rho\sigma\tau} + \phi_{;\sigma\rho\tau} = (-\phi_{;\rho\sigma} + \phi_{;\sigma\rho})_{;\tau} = 0.$$

Since the second covariant derivative of a vector is not necessarily symmetric,

$$\phi_{;\rho\sigma\tau} - \phi_{;\rho\tau\sigma} = (\phi_{;\rho})_{;\sigma\tau} - (\phi_{;\rho})_{;\tau\sigma} = \phi_{;\alpha} R^{\alpha}{}_{\rho\sigma\tau}.$$

In the last two equations, permute the suffixes cyclically ($\rho\sigma\tau \to \sigma\tau\rho \to \tau\rho\sigma$) to yield a total of *six* equations, and add these equations together. The twelve terms on the left sides cancel in pairs, leaving

$$0 = \phi_{;\alpha}(R^{\alpha}{}_{\rho\sigma\tau} + R^{\alpha}{}_{\sigma\tau\rho} + R^{\alpha}{}_{\tau\rho\sigma}).$$

Since ϕ is *arbitrary*, there follows the

$$\textbf{cyclic symmetry:}\quad R^{\alpha}{}_{\rho\sigma\tau} + R^{\alpha}{}_{\sigma\tau\rho} + R^{\alpha}{}_{\tau\rho\sigma} \equiv 0. \tag{6.10}$$

The next symmetry is really a property of the metric tensor. Since the covariant derivative of the metric tensor is identically zero everywhere, so is the second covariant derivative, whence

$$0 \equiv g_{\mu\nu;\rho\sigma} - g_{\mu\nu;\sigma\rho} = g_{\alpha\nu} R^{\alpha}{}_{\mu\rho\sigma} + g_{\mu\alpha} R^{\alpha}{}_{\nu\rho\sigma}. \tag{6.11}$$

(Note that *each* suffix μ, ν generates an R-term in the style of (6.7).) Let us write the 'unmixed' version of the Riemann tensor as

$$R_{\mu\nu\rho\sigma} = g_{\mu\alpha} R^{\alpha}{}_{\nu\rho\sigma}$$

where the superfix has been lowered to become the *first* of the suffixes (regrettably, there is little agreement as to where it should be put). Then (6.11) becomes

$$\textbf{antisymmetry in the first pair of suffixes:} \quad R_{\nu\mu\rho\sigma} + R_{\mu\nu\rho\sigma} \equiv 0. \tag{6.12}$$

All the other symmetries may be obtained from these. Easily the most important is **symmetry on interchange of the first pair of suffixes and the second**:

$$R_{\mu\nu\rho\sigma} \equiv R_{\rho\sigma\mu\nu}. \tag{6.13}$$

(Of course, the order in each pair must be kept the same. See Problem 3.)

6.5 Curvature can get quite complicated!

The Riemann tensor satisfies no further obligatory symmetries. (We shall take this important fact for granted.) It is of interest to ask how many independent elements the Riemann tensor has at an arbitrary point in an nD space; a count based on the symmetries needs to be carried out.

In n dimensions, $R_{\mu\nu\rho\sigma}$ has n^4 elements. On account of the antisymmetries (6.9) and (6.12) many of these are zero. To obtain a nonzero element we may choose μ and ρ arbitrarily (n ways each), but then are somewhat restricted in our choice of ν and σ ($n - 1$ ways each). There may therefore be as many as $n^2(n-1)^2$ nonzero elements.

Not all of these elements are independent. Consideration of the remaining symmetries reveals that there are in total $n^2(n^2-1)/12$ independent elements (Problem 4). This number is a measure of how complicated curvature may become in n-dimensional space. The first few values are

n	$n^2(n^2-1)/12$	
1	0	Trivial
2	1	Easy
3	6	
4	20	Difficult
5	50	
6	105	
7	196	
8	336	Utterly awful
9	540	
10	825	

Many of the illustrative examples of curvature in this book are 2D, and therefore 'easy'. The general problem of 4D looks to be 20 times more difficult, and it is not therefore surprising that progress over the 40 years after 1915 was so slow. (A 'difficulty parameter' is not easy to define. I suspect that a better measure might be, not 20, but e^{20}.)

6.6 Calculating the Riemann tensor

The Riemann tensor in 4D has 256 elements and each element is defined by an expression with ten terms (when the summations are accounted for). Unless we can think of a labour-saving method, its calculation is likely to be tedious, even though we are aware that only 20 independent elements need to be found. One helpful technique goes as follows.

The definition of the Riemann tensor has been given earlier. Let us collect the 256 elements into 16 sets of 16 elements each as follows: define sixteen 4×4 matrices $\mathbf{B}$, one for each pair of values of $\rho\sigma$,

$$R^{\alpha}{}_{\beta\rho\sigma} = \text{the element in row } \alpha \text{ and column } \beta \text{ of } \mathbf{B}_{\rho\sigma}.$$

Then an equivalent definition of the Riemann tensor is (at a rate of sixteen elements at a time)

$$\mathbf{B}_{\rho\sigma} = \partial_\rho \mathbf{\Gamma}_\sigma - \partial_\sigma \mathbf{\Gamma}_\rho + \mathbf{\Gamma}_\rho \mathbf{\Gamma}_\sigma - \mathbf{\Gamma}_\sigma \mathbf{\Gamma}_\rho, \tag{6.14}$$

where the four matrices $\mathbf{\Gamma}$ are exactly those which have already been introduced in connection with parallel transport. Hence the recipe:

(i) Assemble the four matrices $\mathbf{\Gamma}_\rho$, as in section 5.7;
(ii) Calculate six matrices $\mathbf{B}_{\rho\sigma}$ according to the above recipe.

(*Six*, not sixteen, on account of the evident antisymmetry in ρ and σ. Each calculation will use one of the six ways of selecting a pair of different $\mathbf{\Gamma}$-matrices.) It should be clear that much of the strain of collecting the correct combinations of Γs is borne by the matrix multiplications.

There are other recipes; this is the one I happen to prefer for working by hand. Serious work is done by computer.

6.7 An example

Later we shall need to calculate the Riemann tensor for metrics in 4D. Here we provide a simple 2D example to illustrate the method.

The metric for a sphere of radius R is the familiar

$$ds^2 = R^2(d\theta^2 + \sin^2\theta \, d\phi^2).$$

The three nonzero Christoffel symbols have been obtained already in section 5.3. They lead to the matrices

$$\mathbf{\Gamma}_\theta = \begin{bmatrix} \cdot & \cdot \\ \cdot & \cot\theta \end{bmatrix}, \quad \mathbf{\Gamma}_\phi = \begin{bmatrix} \cdot & -\sin\theta\cos\theta \\ \cot\theta & \cdot \end{bmatrix}.$$

The four matrices embodying the 16 elements of the Riemann tensor are $\mathbf{B}_{\theta\theta}$, $\mathbf{B}_{\theta\phi}$, $\mathbf{B}_{\phi\theta}$, and $\mathbf{B}_{\phi\phi}$. However, on account of the evident antisymmetry, only one needs to be evaluated. Let us look at $\mathbf{B}_{\theta\phi}$; the separate terms are

$$\partial_\theta \mathbf{\Gamma}_\phi = \begin{bmatrix} \cdot & -\cos^2\theta + \sin^2\theta \\ -\operatorname{cosec}^2\theta & \cdot \end{bmatrix}, \quad -\partial_\phi \mathbf{\Gamma}_\theta = 0 \text{ (evidently)},$$

$$\mathbf{\Gamma}_\theta \mathbf{\Gamma}_\phi = \begin{bmatrix} \cdot & \cdot \\ \cot^2\theta & \cdot \end{bmatrix}, \quad -\mathbf{\Gamma}_\phi \mathbf{\Gamma}_\theta = \begin{bmatrix} \cdot & \cos^2\theta \\ \cdot & \cdot \end{bmatrix}.$$

The sum of the four contributions is

$$\mathbf{B}_{\theta\phi} = \begin{bmatrix} \cdot & \sin^2\theta \\ -1 & \cdot \end{bmatrix}, \tag{6.15}$$

from which four elements of the Riemann tensor may be read off,

$$R^{\theta}{}_{\theta\theta\phi}=0, \quad R^{\theta}{}_{\phi\theta\phi}=\sin^2\theta, \quad R^{\phi}{}_{\theta\theta\phi}=-1, \quad R^{\phi}{}_{\phi\theta\phi}=0. \tag{6.16}$$

6.8 Principal curvatures

Write

$$R^{\mu\nu}{}_{\rho\sigma}=g^{\nu\alpha}R^{\mu}{}_{\alpha\rho\sigma}, \tag{6.17}$$

raising the first of the suffixes. We use this tensor in the following way: from any antisymmetric tensor $N^{\rho\sigma}$ ($\equiv -N^{\sigma\rho}$), we may form a further antisymmetric tensor

$$R^{\mu\nu}{}_{\rho\sigma}N^{\rho\sigma};$$

this immediately suggests that we consider the *eigenvalue equation*

$$\tfrac{1}{2}R^{\mu\nu}{}_{\rho\sigma}N^{\rho\sigma}=\kappa N^{\mu\nu}, \tag{6.18}$$

where κ is a scalar. This makes good tensor sense: all the affixes are in the right places.

It also makes good *algebraic* sense: we have here a set of effectively six linear homogeneous equations in effectively six unknowns $\mathbf{N}$. (Six, not sixteen, on account of the antisymmetry. This is for 4D; in the nD case the number is $n(n-1)/2$.) In general, such a set of equations has only the trivial solution $\mathbf{N}=0$; exceptions occur when the determinant of the coefficients is zero. This determinant is a polynomial of degree six in κ; its six zeros are the six **principal curvatures**.

Being scalars, the principal curvatures are independent of the coordinate system. They therefore express very clearly the magnitude of the curvature at any point. (Physically, they provide an unambiguous estimate of the strength of the gravity field in the vicinity.) In particular, they may reveal whether a singularity in the metric tensor $\mathbf{g}$ arises from a genuine singularity in space, or whether it is there simply on account of an unfortunate choice of coordinates.

To each of the six roots κ there corresponds a different 'eigensolution' $\mathbf{N}$. These six antisymmetric tensors may be related to each other in a variety of recondite ways, various specializations leading to the different **Petrov–Pirani types** of space.

6.9 The example continued

The best version of the Riemann tensor for giving an estimate for the actual magnitude of the curvature of a curved space is the one most closely associated with the scalar principal curvatures, namely $R^{\mu\nu}{}_{\rho\sigma}$. The values of the elements in this version are the entries in the matrices

$$\mathbf{C}_{\rho\sigma}=\mathbf{B}_{\rho\sigma}\mathbf{g}^{-1}, \tag{6.19}$$

where $\mathbf{g}^{-1}$ is the matrix whose entries are $g^{\mu\nu}$, arranged in the obvious way.

In the example,

$$\mathbf{g}^{-1}=\begin{bmatrix} R^{-2} & \cdot \\ \cdot & R^{-2}\sin^{-2}\theta \end{bmatrix},$$

from which it follows

$$\mathbf{C}_{\theta\phi}=\begin{bmatrix} \cdot & R^{-2} \\ -R^{-2} & \cdot \end{bmatrix}.$$

In two dimensions, there is only one principal curvature, which happens to be the element $\kappa = R^{12}{}_{12}$. Thus in this example, we find that the sphere is a surface of *constant positive curvature* R^{-2}. This is in full accord with the result of Problem 5, Chapter 5.

6—Notes and problems

1. (i) For Cartesian coordinates x, y in the plane, the Christoffel symbols are all zero. The four components of the covariant derivative of a vector field therefore reduce to the ordinary derivatives:

$$V^x{}_{;x} = \partial_x V^x, \quad V^y{}_{;x} = \partial_x V^y, \quad V^x{}_{;y} = \partial_y V^x, \quad V^y{}_{;y} = \partial_y V^y.$$

(ii) For polar coordinates r, θ in the plane, three of the Christoffel symbols are not zero; see section 5.8. Show that the covariant derivative of a vector field has as its components

$$V^r{}_{;r} = \partial_r V^r, \qquad V^r{}_{;\theta} = \partial_\theta V^r - rV^\theta,$$

$$V^\theta{}_{;r} = \partial_r V^\theta + \frac{1}{r}V^\theta, \quad V^\theta{}_{;\theta} = \partial_\theta V^\theta + \frac{1}{r}V^r.$$

2. A vector field in the plane has constant unit length and constant direction parallel to the x-axis. Show that its contravariant components are

$$\text{(in Cartesian coordinates)} \quad V^x = 1, \qquad V^y = 0;$$

$$\text{(in polar coordinates)} \quad V^r = \cos\theta, \quad V^\theta = -\frac{\sin\theta}{r}.$$

Evaluate the covariant derivative in each coordinate system, showing that the result is zero in both cases.

3. One instance of the cyclic symmetry is

$$R_{1234} + R_{1342} + R_{1423} = 0.$$

Obtain three other instances by permuting the indices cyclicly: $1234 \to 2341 \to 3412 \to 4123$, the first being

$$R_{2341} + R_{2413} + R_{2134} = 0.$$

Add the four instances, showing that eight of the twelve terms on the left cancel with the help of the antisymmetries, and that the remaining four are

$$R_{1342} + R_{2413} + R_{3124} + R_{4231} = 0.$$

Again use the antisymmetries to reduce this to

$$R_{1342} = R_{4213},$$

an instance of the further symmetry (6.13).

4. **How many independent elements does the Riemann tensor have?** For the moment we leave the cyclic symmetry on one side. The remaining symmetries may be

used to classify the nonzero elements into mutually independent groupings. Different types of grouping are

Typical instance	Number of such groupings	Number of elements thus accounted for
Class 1		
$R_{1212} = -R_{2112} = -R_{1221} = R_{2121}$	$n(n-1)/2$	$2n(n-1)$
Class 2		
$R_{1213} = -R_{2113} = -R_{1231} = R_{2131}$ $= R_{1312} = -R_{1321} = -R_{3112} = R_{3121}$	$n(n-1)(n-2)/2$	$4n(n-1)(n-2)$
Class 3		
$R_{1234} = -R_{2134} = -R_{1243} = R_{2143}$ $= R_{3412} = -R_{3421} = -R_{4312} = R_{4321}$	$n(n-1)(n-2)(n-3)/8$	$n(n-1)(n-2)(n-3)$
		Total $= n^2(n-1)^2$

The cyclic symmetry tells us nothing new as far as the first two classes are concerned. But in (for example)

$$R_{1234} + R_{1342} + R_{1423} = 0$$

the three elements on the left belong to three distinct groupings in class 3; knowing the elements in two of the groupings gives us the elements in the third. Thus only 2/3 of the groupings in class 3 are independent.

The total number of independent groupings gives the total number of independent elements in the Riemann tensor. This is

$$\begin{aligned} &n(n-1)/2 + n(n-1)(n-2)/2 + (2/3)n(n-1)(n-2)(n-3)/8 \\ &\quad = n^2(n^2-1)/12. \end{aligned}$$

5. We are now in a position to undertake a solution to Problem 1 of Chapter 3 (see also Problem 2 of Chapter 5). For each of the metrics, evaluate the Christoffel symbols, use the techniques of this chapter (or any technique you may prefer) to find the Riemann tensor, and hence show that in each case the curvature is uniform.

   [You ought to find

$$\text{(i) } R^{r\phi}{}_{r\phi} = 0; \quad \text{(ii) } R^{\theta\phi}{}_{\theta\phi} = 1; \quad \text{(iii) } R^{uv}{}_{uv} = 0;$$
$$\text{(iv) } R^{r\phi}{}_{r\phi} = 1; \quad \text{(v) } R^{xy}{}_{xy} = 0; \quad \text{(vi) } R^{w\phi}{}_{w\phi} = -1. \quad]$$

6. Later we shall need the metric for a 3D space of constant curvature. One version is

$$ds^2 = \frac{dr^2}{1-kr^2} + r^2\, d\theta^2 + r^2 \sin^2\theta\, d\phi^2$$

where k is constant.

   Ten of the 27 Christoffel symbols are nonzero. Find them, and show that they lead to the customary matrices

$$\boldsymbol{\Gamma}_r = \begin{bmatrix} kr/(1-kr^2) & \cdot & \cdot \\ \cdot & 1/r & \cdot \\ \cdot & \cdot & 1/r \end{bmatrix};$$

$$\boldsymbol{\Gamma}_\theta = \begin{bmatrix} \cdot & -r(1-kr^2) & \cdot \\ 1/r & \cdot & \cdot \\ \cdot & \cdot & \cot\theta \end{bmatrix};$$

$$\boldsymbol{\Gamma}_\phi = \begin{bmatrix} \cdot & \cdot & -r(1-kr^2)\sin^2\theta \\ \cdot & \cdot & -\sin\theta\cos\theta \\ 1/r & \cot\theta & \cdot \end{bmatrix}.$$

Evaluate the matrices **C** (three will be sufficient, on account of the antisymmetries), and show that each element takes on of the three values 0, $+k$, $-k$.

Show that the three principal curvatures are all k. (The space is isotropic, homogeneous, and of constant curvature k.)

7. (i) In the metric of the previous problem, setting the curvature k to zero leads to the metric for 'ordinary' 3D space in spherical polar coordinates.
(ii) If k is *positive*, introduce a new radial coordinate χ by

$$r = \frac{\sin\chi}{\sqrt{k}}$$

to obtain the metric

$$\mathrm{d}s^2 = R^2(\mathrm{d}\chi^2 + \sin^2\chi\,\mathrm{d}\theta^2 + \sin^2\chi\sin^2\theta\,\mathrm{d}\phi^2)$$

in which $R^2 = 1/k$.
(iii) If k is *negative*, introduce the new radial coordinate χ by

$$r = \frac{\sinh\chi}{\sqrt{-k}}$$

to obtain the metric

$$\mathrm{d}s^2 = (1/|k|)(\mathrm{d}\chi^2 + \sinh^2\chi\,\mathrm{d}\theta^2 + \sinh^2\chi\sin^2\theta\,\mathrm{d}\phi^2).$$

In these versions of the metric, the coordinate χ is strictly proportional to distance measured radially from the origin at $\chi = 0$. This will make their use particularly convenient when we come to consider cosmology in Chapter 11.

8. When k is positive in the metric of Problem 6, the range of r is limited by $kr^2 < 1$. An apparently identical metric, in which the range of r satisfies $kr^2 > 1$, is

$$\mathrm{d}s^2 = -\frac{\mathrm{d}r^2}{kr^2 - 1} + r^2\,\mathrm{d}\theta^2 + r^2\sin^2\theta\,\mathrm{d}\phi^2.$$

This metric belongs to a (2 + 1)D space of constant curvature k. The two spaces are not connected in any useful way whatever. The fact that one range of r seems to lead naturally into the other is misleading. It arises from the singularity in **g** when $kr^2 = 1$; at this value, the metric is essentially *ambiguous*. A different choice of coordinates in either case may be used to remove the ambiguity.

9. What actually happens at $kr^2 = 1$? After the change of coordinate given in Problem 7(ii), the awkward value of r corresponds to $\chi = \pi/2$; when written in terms of χ the metric does not show any singular behaviour whatever. When written in terms of r, the metric acquires a singularity at the point where $\sin\chi$ has a maximum value. At this point, the coordinate r is incapable of representing the space in any sensible way, and the singularity is the fault of the coordinate system. It is not 'really' there in the space itself.

   We shall need to face an exactly similar problem when we look at black holes in the light of the Schwarzschild metric in Chapter 9.

10. In the metric of Problem 6, change to a new radial coordinate R by $r = R/(1 + kR^2/4)$, to obtain

$$ds^2 = \frac{dR^2 + R^2\,d\theta^2 + R^2\sin^2\theta\,d\phi^2}{(1 + kR^2/4)^2}.$$

Note that the numerator is the metric for a flat space in polar coordinates $r\theta\phi$. Any such metric which, apart from an overall multiplying function of position, is the metric of a flat space is called **conformal**.

We can introduce *pseudo-Cartesian coordinates* by the familiar relations

$$X = R\sin\theta\cos\phi, \quad Y = R\sin\theta\sin\phi, \quad Z = R\cos\theta.$$

Show that the metric becomes

$$ds^2 = \frac{dX^2 + dY^2 + dZ^2}{\{1 + k(X^2 + Y^2 + Z^2)/4\}^2}.$$

This is yet another representation of 3D space of uniform curvature k. When $k = 0$, we recover Pythagoras' theorem for 3D Euclidean space.

On account of the evident symmetry between X and Y and Z, this form is ideal for certain kinds of calculation. On account of the nature of the coordinate χ in Problem 7, the equivalent metrics given there are ideal for other kinds. We must be flexible!

7

But what exactly *is* curvature?

We have said much about curvature and given several hints as to its nature. Some recapitulation and consolidation is now called for, followed by a formal definition. This chapter deals briefly with some of the mathematical aspects, while the physical manifestation of curvature will be considered in the next.

7.1 Extrinsic and intrinsic curvature

Imagine, if you will, certain surfaces in ordinary 3D space: a plane, a circular cylinder, and a sphere.

The plane is *flat*; its curvature is everywhere zero. We take this to be evident. The cylinder is equally evidently *curved*: it has a radius of curvature which may be straightforwardly measured in the 3D space. It is important to recognize the nature of this curvature: the cylinder is not curved 'in itself'; it is curved as a result of the way it is embedded in a space of higher dimension. The evidence is that it may be constructed from a sheet of flat cardboard, without stretching or tearing, simply by rolling it up. We say that its curvature is **extrinsic** only.

It is different with the sphere. Once again, the sphere is manifestly curved extrinsically on account of its embedding. Additionally, however, it cannot be formed from a sheet of cardboard without deformation; its **intrinsic** geometry—the geometry of relationships and diagrams *on* the surface—differs from the intrinsic (Euclidean) geometry of the plane. (This causes much trouble to atlas-makers.) This is a property of the surface itself, and has little to do with the embedding: the sphere has **intrinsic** curvature.

We may go one small step further. Forget about the 3D space, and about the embedding-related extrinsic curvature with its concomitant radii of curvature, and think of the surfaces in isolation. Intrinsic curvature is all that remains with any meaning; in some sense it is a property of the geometry in the surface alone. In fact, it *is* this geometry.

When we talk about the curvature of a worldline (its departure from straightness—its acceleration), we are of course referring to its *extrinsic* curvature, which has to do with how it is embedded in spacetime and with its relationship to its environment. By contrast, when we talk about the curvature of spacetime itself, we must resist any temptation to think of it as embedded in a 'higher' space, because there isn't one. In this case, therefore, we must consider intrinsic curvature alone, and search for it in the internal geometry of spacetime itself.

7.2 The essential nature of intrinsic curvature

Intrinsic curvature is defined with the help of parallel transport. We have already seen that parallel transport along a curve effectively entails a linear transformation (a rotation, or perhaps a Lorentz transformation, depending on the signature of the space; see section 5.7.) Now what if the curve returns to its starting point, forming a **circuit**? There is no guarantee that the rotation returns to its starting orientation. *Any change in orientation is a measure of the total intrinsic curvature embraced by the circuit.*

The curvature associated with a circuit is not just a single number (except, as it happens, in 2D); it is a collection of numbers large enough to specify a change in orientation. It is a complicated concept.

It is useful to be able to discuss the curvature *at a particular point* P. To do this, we need to decide

(i) how to represent a *small* circuit in the tensor formalism; and
(ii) how to calculate the effect of parallel transport round such a small circuit.

The curvature at a point P in the space will then be known from the effects of parallel transport round a large enough selection of small circuits in the immediate vicinity of P.

7.3 Infinitesimal circuits

An **infinitesimal circuit** is a small, closed path in the immediate neighbourhood of a point P; it is conveniently described parametrically by functions $z^{\mu}(\sigma)$ as σ goes from 0 to 1, as shown in Fig. 7.1. The **tensor area** of the circuit is defined as

$$Q^{\alpha\beta} = \int_0^1 z^{\alpha} \frac{\mathrm{d}z^{\beta}}{\mathrm{d}\sigma} \mathrm{d}\sigma \equiv \int_{\mathrm{C}} z^{\alpha}\, \mathrm{d}z^{\beta}; \tag{7.1}$$

this gives not only the magnitude of the area, but also its orientation in a very natural way. Integration by parts shows that $\mathbf{Q}$ is always *antisymmetric*:

$$Q^{\alpha\beta} = -Q^{\beta\alpha}.$$

To see how reasonable this definition is, evaluate $\mathbf{Q}$ for the small triangle shown in Fig. 7.2; the result is

$$Q^{\alpha\beta} = \tfrac{1}{2}(a^{\alpha}b^{\beta} - a^{\beta}b^{\alpha})$$

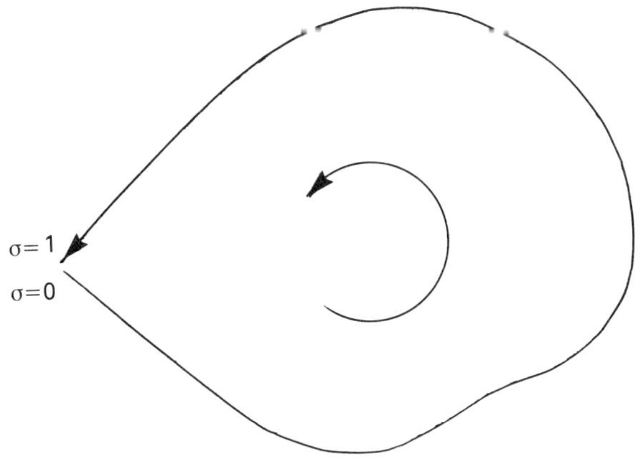

Fig. 7.1

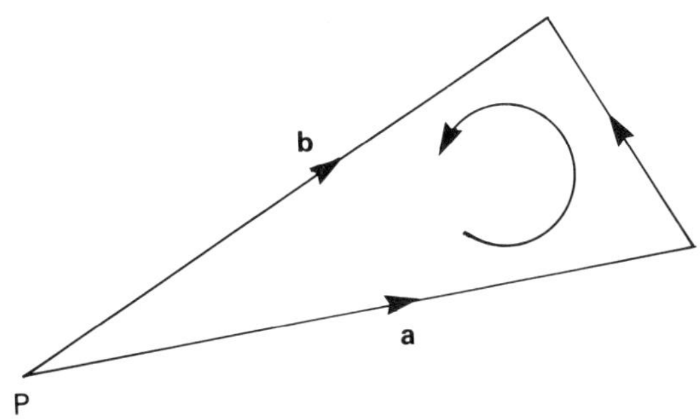

Fig. 7.2

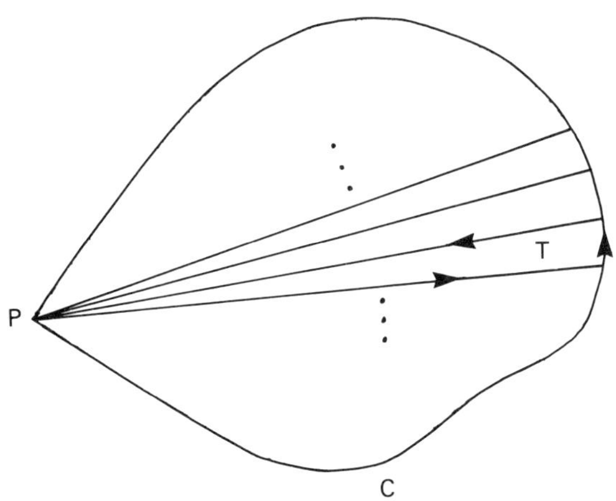

Fig. 7.3

by a simple evaluation (Problem 1). The 'orientation' of $\mathbf{Q}$ is determined by the orientations of the vectors **a** and **b**, which clearly determine the plane of the triangle. The 'magnitude' $|Q|$ is defined by

$$\begin{aligned}|Q|^2 = \tfrac{1}{2}Q^{\alpha\beta}Q_{\alpha\beta} &= \tfrac{1}{8}(a^\alpha b^\beta - a^\beta b^\alpha)(a_\alpha b_\beta - a_\beta b_\alpha)\\ &= \tfrac{1}{4}(a^2b^2 - (\mathbf{a}\cdot\mathbf{b})^2)\\ &= \tfrac{1}{4}a^2b^2\sin^2\theta\end{aligned}$$

where θ is the angle between **a** and **b**. Thus $|Q|$ is exactly the area of the triangle. **Q** therefore describes an infinitesimal triangle rather well, both as to area and as to orientation.

More complicated circuits may be triangulated in a kind of daisy pattern as shown in Fig. 7.3. The tensor area for the circuit C is then just the sum of the tensor areas for the individual triangles T, since the contributions from the spokes emanating from P clearly cancel in the evaluation of the integral.

7.4 Parallel transport round an infinitesimal circuit

The aim now is to integrate the equation (5.9) for parallel transport,

$$\dot{\mathbf{T}}(\sigma) = -\mathbf{M}(\sigma)\mathbf{T}(\sigma)$$

where

$$\mathbf{M}(\sigma) = \dot{x}^\alpha \mathbf{\Gamma}_\alpha$$

once round an infinitesimal circuit **Q**. If the space is curved, we do not expect **T** to return to its original value **I** on completing the circuit; the discrepancy will be used to define the curvature of the space at that locality. From our experience of spherical excess (Problem 5 of Chapter 5) we know that we shall need to work to orders of smallness where the area of the circuit is significant, that is to second order in **z**, or equivalently to second order in **M**.

To this order, the parallel-transport equation is satisfied by

$$\mathbf{T}(\sigma) = \mathbf{I} - \int_0^\sigma \mathbf{M}(\sigma_1)\,\mathrm{d}\sigma_1 + \int_0^\sigma \mathbf{M}(\sigma_1)\,\mathrm{d}\sigma_1 \int_0^{\sigma_1} \mathbf{M}(\sigma_2)\,\mathrm{d}\sigma_2 + O(\mathbf{M}^3)$$

as may be checked by direct substitution. (Note that the second-order term involves a matrix multiplication, and the order of the factors is significant.) We are interested in $\mathbf{T}(1)$.

Now Taylor expansion gives

$$\mathbf{\Gamma}_\alpha(\sigma) = \mathbf{\Gamma}_\alpha^{\text{at P}} + z^\beta[\partial_\beta \mathbf{\Gamma}_\alpha]^{\text{at P}} + O(z^2).$$

Insertion into the expression for **T** and retention of terms to second order only gives

$$\begin{aligned}-\int_0^1 \mathbf{M}(\sigma) &= -\int_0^1 \dot{z}^\alpha(\mathbf{\Gamma}_\alpha^{\text{at P}} + z^\beta[\partial_\beta\mathbf{\Gamma}_\alpha]^{\text{at P}})\,\mathrm{d}\sigma\\ &= Q^{\alpha\beta}[\partial_\beta\mathbf{\Gamma}_\alpha]^{\text{at P}}\end{aligned}$$

and

$$\begin{aligned}\int_0^1 \mathbf{M}(\sigma_1)\,\mathrm{d}\sigma_1 \int_0^{\sigma_1} \mathbf{M}(\sigma_2)\,\mathrm{d}\sigma_2 &= \int_0^1 \frac{\mathrm{d}z^\alpha}{\mathrm{d}\sigma_1}\mathbf{\Gamma}_\alpha^{\text{at P}}\,\mathrm{d}\sigma_1 \int_0^{\sigma_1} \frac{\mathrm{d}z^\beta}{\mathrm{d}\sigma_2}\mathbf{\Gamma}_\beta^{\text{at P}}\,\mathrm{d}\sigma_2\\ &= -Q^{\alpha\beta}\mathbf{\Gamma}_\alpha^{\text{at P}}\mathbf{\Gamma}_\beta^{\text{at P}}.\end{aligned}$$

(Everything with a superfix 'at P' is evaluated at the fixed point P; from now on we drop the P.) Finally

$$\mathbf{T}(1) = \mathbf{I} - (1/2)Q^{\alpha\beta}\mathbf{B}_{\alpha\beta} + O(z^3), \tag{7.2}$$

where, since $\mathbf{Q}$ is antisymmetric, we are able to write

$$\mathbf{B}_{\alpha\beta} = \partial_\alpha \mathbf{\Gamma}_\beta - \partial_\beta \mathbf{\Gamma}_\alpha + \mathbf{\Gamma}_\alpha \mathbf{\Gamma}_\beta + \mathbf{\Gamma}_\beta \mathbf{\Gamma}_\alpha. \tag{7.3}$$

This is of course immediately recognized as the Riemann tensor of (6.14) which therefore describes the curvature at P.

7.5 Is this all?

Of course, we have evaluated $\mathbf{T}(1)$ to second order only, and we may ask whether higher order terms may involve *new* tensors giving further information about the curvature. In fact the answer is negative, on account of a fundamental theorem: *any tensor which depends only on the metric tensor can be written as a tensor expression involving the metric tensor, the Riemann tensor, and the covariant derivatives of the Riemann tensor alone*. Thus nothing essentially new is to be expected at higher orders. We therefore assert that the Riemann tensor tells us all that there is to be known about the intrinsic curvature of the space; in a real sense, it *is* this curvature.

An informal way to see that the Riemann tensor is all we need to understand curvature is to think about parallel transport round any large circuit (that is, one that is not infinitesimal). Regard this circuit as the boundary of some surface bridging the intervening space, and cover this surface with a network of small meshes (infinitesimal circuits!). We may easily believe that parallel transport round the finite circuit can in some way be expressed as a composition of the contributions from each mesh, given to high enough order by $\mathbf{I} - (1/2)Q^{\alpha\beta}\mathbf{B}_{\alpha\beta}$. The details in more than 2D are very troublesome, being complicated by noncommutativity of the matrices, and a consequent need to choose the mesh contours rather carefully. But the fact that it can be done at all implies that the Riemann tensor says everything.

By the way, if this procedure can be done for one surface bridging the circuit, then it can be done for any other from the enormous choice of such surfaces. In order that *any* choice should give the same result for parallel transport round the circuit—as it obviously must—the Riemann tensor must satisfy some differential identity or other. The identity in question is the **Bianchi identity**; it has an important consequence for the physics, and we shall need to return to it in Chapter 10. (See also problem 3.)

7.6 Curves parallel to a geodesic

Imagine a geodesic G and a neighbouring **parallel curve** P generated from G as follows. A small displacement $\boldsymbol{\lambda}$ is parallel-transported along G: its extremity traces out P (Fig. 7.4). Clearly, P is not expected to be a geodesic unless the space is flat. For example, the equator of the Earth is a geodesic, while the parallel curve joining all points at latitude 1 °N clearly is not.

We aim to compute the curvature (or acceleration, if it is spacetime we are dealing with) of a curve running parallel to a nearby geodesic. Suppose A, B, C are three neighbouring points on G such that BC results from the parallel transport of AB along itself (this is the characteristic property of a geodesic). The points A′, B′, C′ on

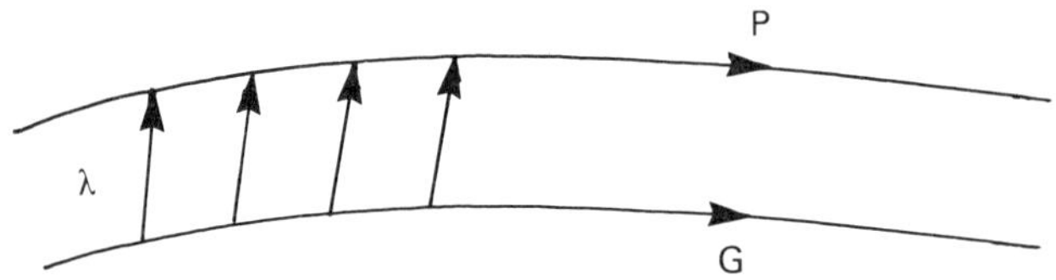

Fig. 7.4

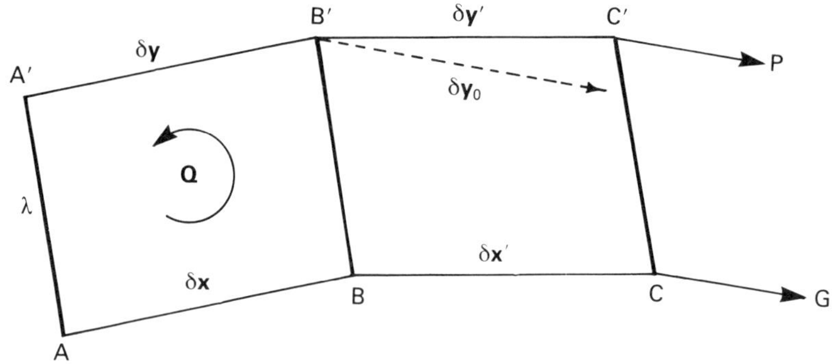

Fig. 7.5

the parallel curve arise from the parallel transport of $\boldsymbol{\lambda}$ along $A \rightarrow B$ and $B \rightarrow C$. We end up with two adjacent infinitesimal parallelograms as shown in Fig. 7.5.

To find the curvature of P we need to determine by how much $\delta \mathbf{y}'$ fails to be the parallel transport of $\delta \mathbf{y}$ along $A' \rightarrow B'$. Consider the parallel transport of $\delta \mathbf{y}$ along the more indirect three-step route $A' \rightarrow A \rightarrow B \rightarrow B'$; the consecutive changes are

$\delta \mathbf{y} \rightarrow \delta \mathbf{x}$	as	$A' \rightarrow A$	(opposite sides of a parallelogram)
$\rightarrow \delta \mathbf{x}'$	as	$A \rightarrow B$	(G is a geodesic)
$\rightarrow \delta \mathbf{y}'$	as	$B \rightarrow B'$	(opposite sides of a parallelogram).

This result is to be compared with the result of the parallel transport of $\delta \mathbf{y}$ *directly* along $A' \rightarrow B'$, $\delta \mathbf{y}_0$ say, which may not be the same as $\delta \mathbf{y}'$ if the space is curved; the discrepancy is $\delta \mathbf{y}' - \delta \mathbf{y}_0$, and will be the result of parallel transport round the circuit $A' \rightarrow A \rightarrow B \rightarrow B' \rightarrow A'$. The tensor area of this circuit is

$$Q^{\alpha\beta} = \delta x^\alpha \lambda^\beta - \delta x^\beta \lambda^\alpha.$$

Hence the discrepancy in $\delta \mathbf{y}$ is

$$-\tfrac{1}{2} Q^{\alpha\beta} \mathbf{B}_{\alpha\beta} \delta \mathbf{y},$$

or, more explicitly,

$$-\delta x^\alpha \lambda^\beta R^\mu{}_{\nu\alpha\beta} \delta x^\nu$$

($\delta \mathbf{x}$ will do instead of $\delta \mathbf{y}$, to the order of accuracy required). Dividing through by δs^2 and allowing $\delta s \rightarrow 0$ yields the curvature of the curve P

$$F^\mu = -\dot{x}^\alpha \dot{x}^\nu R^\mu{}_{\nu\alpha\beta} \lambda^\beta. \tag{7.4}$$

Setting $\boldsymbol{\lambda} = 0$ of course gives the geodesic itself, along which $\mathbf{F} = 0$.

Thinking about the collection of neighbouring worldlines parallel to a geodesic worldline will help us to formulate the correct field equations for gravity in empty space.

7—Notes and problems

1. The tensor area of a small triangle OAB may be found as follows. Describe the three sides parametrically by

$$\begin{array}{lll} \mathrm{OA}: \mathbf{z} = \lambda\mathbf{a} & \mathrm{d}\mathbf{z} = \mathbf{a}\,\mathrm{d}\lambda & 0 < \lambda < 1, \\ \mathrm{AB}: \mathbf{z} = \mathbf{a} + \mu(\mathbf{b} - \mathbf{a}) & \mathrm{d}\mathbf{z} = (\mathbf{b} - \mathbf{a})\,\mathrm{d}\mu & 0 < \mu < 1, \\ \mathrm{BO}: \mathbf{z} = (1 - \nu)\mathbf{b} & \mathrm{d}\mathbf{z} = -\,\mathbf{b}\,\mathrm{d}\nu & 0 < \nu < 1. \end{array}$$

The integral round the circuit falls into three parts, the first being

$$\int_{\mathrm{OA}} z^\alpha\,\mathrm{d}z^\beta = \int_0^1 \lambda a^\alpha \cdot a^\beta\,\mathrm{d}\lambda = \tfrac{1}{2}a^\alpha a^\beta.$$

Evaluate the others, and show that the sum of all three is

$$\tfrac{1}{2}(a^\alpha b^\beta - a^\beta b^\alpha).$$

2. The vectors **a** and **b** are small, and fixed at P. As χ goes from 0 to 2π, the vector $\mathbf{z} = \mathbf{a}\cos\chi + \mathbf{b}\sin\chi$ describes a small ellipse centred on P. Show that

$$Q^{\alpha\beta} = \int z^\alpha\,\mathrm{d}z^\beta = \pi(a^\alpha b^\beta - a^\beta b^\alpha).$$

Evaluate $|Q|$, and show that, if $|a| = |b| = R$ and $\mathbf{a}\cdot\mathbf{b} = 0$, then $|Q| = \pi R^2$. Interpret this result.

3. **The Bianchi identity** In the style of (6.7), verify the two equations (actually 512 equations in 4D):

$$(a_{\mu;\rho\sigma} - a_{\mu;\sigma\rho})_{;\tau} = a_\omega R^\omega{}_{\mu\rho\sigma;\tau} + a_{\omega;\tau} R^\omega{}_{\mu\rho\sigma}$$

and

$$-(a_{\mu;\tau})_{;\rho\sigma} + (a_{\mu;\tau})_{;\sigma\rho} = -\,a_{\omega;\tau} R^\omega{}_{\mu\rho\sigma} - a_{\mu;\omega} R^\omega{}_{\tau\rho\sigma}.$$

Obtain four further equations by permuting $\rho\sigma\tau$ cyclicly: $\rho\sigma\tau \to \sigma\tau\rho \to \tau\rho\sigma$. Add all six equations. Show that all twelve terms on the left, and a certain six on the right, cancel in pairs. Show that a further three cancel on the right, by the cyclic symmetry. From the residue, deduce the Bianchi identity

$$R^\omega{}_{\mu\rho\sigma;\tau} + R^\omega{}_{\mu\sigma\tau;\rho} + R^\omega{}_{\mu\tau\rho;\sigma} = 0.$$

How many equations in 4D does this stand for?

4. On the surface of a sphere of radius R, as s varies, the prescription

$$\theta = \pi/2, \quad \phi = s/R$$

traces out the Equator. The parallel curve at a small latitude ε to the south is traced out by

$$\theta = \pi/2 + \varepsilon, \quad \phi = s/R,$$

so that the vector λ is

$$\lambda^\theta = \varepsilon, \quad \lambda^\phi = 0.$$

Show that in the calculation of **F** according to (7.4) only one term survives,

$$F^\theta = -\dot{\phi}\dot{\phi}R^\theta{}_{\phi\phi\theta}\lambda^\theta = \frac{\varepsilon}{R^2}.$$

Check that this is in full accord with Problem 4 of Chapter 5 when $\theta = \pi/2 + \varepsilon$, to first order in ε.

5. Since a curve parallel to a neighbouring geodesic is not itself expected to be a geodesic in general, a pair of neighbouring geodesics which begin by being parallel will not remain so. Each will appear *curved* with respect to the other; give reasons for supposing that the mutual apparent curvature will be

$$-F^\mu = +\dot{x}^\alpha\dot{x}^\nu R^\mu{}_{\nu\alpha\beta}\lambda^\beta.$$

This is the **geodesic deviation**, used in some texts in preference to parallel curves.

8

The field equations for the curvature of empty spacetime

All the fields of physics satisfy appropriate partial differential equations:

Electromagnetic field	Maxwell's equations
Newtonian gravitational potential	$\nabla^2\phi = 0$ (in empty space)

etc.

This must be true also of the metric tensor; its values at different points in spacetime must be related in some way, otherwise spacetime would be permitted to lash around as it likes! It is to be expected that there shall be some restriction on the curvature of spacetime, which will say how Einstein gravity is to behave, just as $\nabla^2\phi = 0$ says how Newtonian gravity is to behave.

The last remark is the clue to the problem: if we demand that the Einstein and Newton theories become identical in the limit of weak fields—a reasonable demand, since Newtonian gravity comes so close to the truth in, say, the Solar System—then we find that the possibilities for a correct field equation are very few. This chapter is concerned with finding what the possibilities are.

8.1 Tides

In physical spacetime

$$F^\mu = -\dot{x}^\alpha \dot{x}^\nu R^\mu{}_{\nu\alpha\beta}\lambda^\beta \equiv -\Delta^\mu{}_\beta \lambda^\beta \quad \text{(in which } \Delta^\mu{}_\beta \equiv \dot{x}^\alpha \dot{x}^\nu R^\mu{}_{\nu\alpha\beta}\text{)}$$

is the *force per unit mass* (see section 7.6) required to keep a particle moving along a parallel curve, rather than along the neighbouring geodesic. This force must be supplied by some mechanical means or other. For example, the worldline of the centre of mass of a satellite in free fall is a timelike geodesic, but this is not true of other points of the satellite—they have parallel worldlines (if, that is, the satellite is not rotating). Thus the satellite keeps its shape only by virtue of forces within its own structure; from the astronaut's point of view it is as if these forces are needed to balance the **tidal forces**, of gravitational origin. The tidal force per unit mass at the point $\boldsymbol{\lambda}$ (relative to the satellite centre of mass) is therefore just $\Delta^{\mu}{}_{\nu}\lambda^{\nu}$. We shall call $\boldsymbol{\Delta}$ the **tidal tensor**. (The tidal tensor is *symmetric* and *spacelike*: see Problem 1.)

These considerations show how we can measure spacetime curvature by mechanical means: by measuring the stresses in a suitable structure in free fall (a network of spring balances, perhaps) we can find $\boldsymbol{\Delta}$ without trouble. To obtain fuller information about the Riemann tensor we shall need to measure the tidal tensor for a variety of different velocities $\dot{\mathbf{x}}$—to throw our structures around at high speeds in different directions!

An example of a tidal tensor in the specific case of the Schwarzschild metric will be given at the end of this chapter.

8.2 The field equations for empty spacetime

There is no doubt that Newtonian gravity is a remarkably close approximation to the truth, and any competitor (like Einstein's) must not be very different. This gives us an important clue to what the field equations for curvature must be.

The Newtonian gravity field may be described by a potential $\phi(r)$ which yields a (3D) tidal tensor (in Cartesian coordinates)

$$\Delta_{ij}^{\text{Newton}} = -\frac{\partial^2 \phi}{\partial x^i \, \partial x^j} \quad (i, j = 1, 2, 3)$$

(the gravitational acceleration at any point is $-\partial\phi/\partial x^j$ at that point, while the tides are the non-uniformities in gravitational acceleration in the neighbourhood of that point: hence the second derivatives). Moreover, in empty space, the Newtonian potential satisfies Laplace's Equation,

$$\nabla^2 \phi = 0,$$

or, equivalently,

$$\sum_{i=1}^{3} \Delta_{ii}^{\text{Newton}} = 0:$$

the trace of the tidal tensor is zero. In order to have any hope of competing, Einstein gravity must satisfy a similar condition, at least to a high degree of approximation. The obvious course is to demand that

$$\Delta^{\mu}{}_{\mu} = 0 \quad \text{in empty space,} \tag{8.1}$$

i.e.

$$R^{\mu}{}_{\alpha\mu\beta}\dot{x}^{\alpha}\dot{x}^{\beta} = 0 \quad \text{in empty space.}$$

This is a restriction on the **Ricci tensor**

$$R_{\alpha\beta} \equiv R^{\mu}{}_{\alpha\mu\beta}, \tag{8.2}$$

which is easily shown to have the important symmetry

$$R_{\alpha\beta} = R_{\beta\alpha}. \tag{8.3}$$

If the restriction is imposed on the Ricci tensor for *all* velocities (for satellites moving in any direction at any speed), we may immediately conclude that

$$R_{\alpha\beta}=0 \quad \text{in empty space.} \tag{8.4}$$

This requirement gives ten equations (not sixteen, on account of symmetry) to be satisfied by any empty-space metric claiming to have Newtonian gravity as a close approximation.

In fairness it ought to be said that Earth satellites move rather slowly (about $10^{-4}c$ at most), so that the tidal argument is hardly conclusive: all that it shows is that R_{tt} must be nearly zero, and it says little about the other nine. On the other hand, the spirit of relativity is to treat a tensor as a single entity, and having decided to set one component to zero in empty space we can hardly avoid setting all the others to zero too.

This does not quite settle the matter. There is just one slightly more general possibility with any kind of plausibility about it:

$$R_{\alpha\beta}=-\Lambda g_{\alpha\beta} \quad \text{in empty space} \tag{8.5}$$

where Λ is a universal constant. The effect of the extra **cosmological term** is to give spacetime *as a whole* an extra curvature, positive or negative according to the sign of Λ. If Λ is exceedingly small, this extra curvature will have no observable effect within, say, the confines of the Solar System, but may be important over great cosmological distances—just as Earth curvature is irrelevant for snooker table manufacture, but very important for an international airline. The cosmological term was first introduced by Einstein by reason of certain philosophical arguments about the nature of the Universe as a whole. It is now generally held that these arguments were not valid; in any case, astronomical observations do not support the extra term. (Einstein himself later referred to its introduction as perhaps the biggest blunder of his life; the rest of us will feel that such self-denigration is quite uncalled for.)

8.3 Why *four* dimensions?

There is, of course, no answer to this question. However, in General Relativity, (1 + 3)D is the first '*interesting*' number of dimensions. In 3D, the Riemann tensor and the Ricci tensor have the same number (six) of independent components. Therefore it is to be expected that knowing the Ricci tensor will enable one to recover the Riemann tensor. This is in fact the case; *in 3D only*

$$R_{\alpha\beta\gamma\delta}=R_{\alpha\gamma}g_{\beta\delta}+R_{\beta\delta}g_{\alpha\gamma}-R_{\alpha\delta}g_{\beta\gamma}-R_{\beta\gamma}g_{\alpha\delta}-(R_S/2)(g_{\alpha\gamma}g_{\beta\delta}-g_{\alpha\delta}g_{\beta\gamma}) \quad \textbf{(3D only)}$$

in which the **scalar curvature** R_S is defined as

$$R_S=g^{\mu\nu}R_{\mu\nu}$$

(the affix S is there to prevent confusion: the letter R has come to be used for perhaps too many purposes in Relativity). It is clear that

$$R_{\alpha\beta}=0 \text{ implies } R_{\alpha\beta\gamma\delta}=0 \quad \textbf{(3D only)};$$

that is, *empty spacetime has to be flat*. Gravity of the General Relativity kind in (1 + 2)D is very simple: there isn't any.

In 2D the Riemann tensor is even more constrained. Having just one independent component, it may be expressed in terms of the scalar curvature as

$$R_{\alpha\beta\gamma\delta} = (R_S/2)(g_{\alpha\gamma}g_{\beta\delta} - g_{\alpha\delta}g_{\beta\gamma}) \quad \textbf{(2D only)}.$$

Thus in 2D, curvature is essentially *scalar*, making the geometry of curved surfaces very simple in comparison to the geometry of higher dimensions.

8.4 A spherically symmetric static metric for spacetime

Consider the metric

$$c^2\,d\tau^2 = A(r)\,dt^2 - B(r)\,dr^2 - r^2(d\theta^2 + \sin^2\theta\,d\phi^2). \tag{8.6}$$

The coordinates θ and ϕ occur in the combination $r^2(d\theta^2 + \sin^2\theta\,d\phi^2)$, which is the line element for the surface of a sphere of area $4\pi r^2$; they occur nowhere else. The metric is therefore *spherically symmetric*. The coefficients in the metric are all independent of the time coordinate t; the metric is therefore *static*. If also we require that

$$A(r) \to c^2, \quad B(r) \to 1 \quad \text{as } r \to \infty, \tag{8.7}$$

then the metric is asymptotically Minkowskian at large 'distances' r. It is therefore a good candidate for describing the gravitational field of an isolated spherically symmetric static gravitating mass. The coordinate r is not necessarily the radial distance, and there is a considerable arbitrariness in its choice. Here it is defined in such a way that the spherical shell at 'position r' has area $4\pi r^2$. (There is no reason to expect the other coefficients A and B to be independent of r. Indeed, why not have cross-terms like $dt\,dr$ in the metric? More careful arguments show that they can be dispensed with in this case, though they may become essential in other cases. For example, for a *rotating* gravitating mass, a term $dt\,d\phi$ becomes obligatory; however, the angular velocity of objects like the Earth and Sun is so small that the effect of rotation is almost negligible. We shall calculate it later.)

The aim is to determine the functions A and B; here the vacuum case will be examined (the necessary changes in the presence of matter will be dealt with in Chapter 12). The method will be to obtain in turn the Christoffel symbols, the Riemann tensor, and the Ricci tensor; the vacuum Einstein equations Ricci = 0 will then be solved for the functions A and B.

8.5 Calculating the Christoffel symbols

The Lagrangian for the problem is

$$L = A\dot{t}^2 - B\dot{r}^2 - r^2(\dot{\theta}^2 + \sin^2\theta\dot{\phi}^2). \tag{8.8}$$

As usual, the Γs may be read off as coefficients in the geodesic equations. The work exactly parallels that of section 5.12. Once again, it turns out that there are just nine nonzero coefficients, leading to thirteen nonzero Christoffel terms. (We shall write A',

$$\Gamma_t: \begin{bmatrix} \cdot & \frac{A'}{2A} & \cdot & \cdot \\ \frac{A'}{2B} & \cdot & \cdot & \cdot \\ \cdot & \cdot & \cdot & \cdot \\ \cdot & \cdot & \cdot & \cdot \end{bmatrix} \qquad \Gamma_r: \begin{bmatrix} \frac{A'}{2A} & \cdot & \cdot & \cdot \\ \cdot & \frac{B'}{2B} & \cdot & \cdot \\ \cdot & \cdot & \frac{1}{r} & \cdot \\ \cdot & \cdot & \cdot & \frac{1}{r} \end{bmatrix}$$

$$\Gamma_\theta: \begin{bmatrix} \cdot & \cdot & \cdot & \cdot \\ \cdot & \cdot & -\frac{r}{B} & \cdot \\ \cdot & \frac{1}{r} & \cdot & \cdot \\ \cdot & \cdot & \cdot & \cot\theta \end{bmatrix} \qquad \Gamma_\phi: \begin{bmatrix} \cdot & \cdot & \cdot & \cdot \\ \cdot & \cdot & \cdot & \frac{-r\sin^2\theta}{B} \\ \cdot & \cdot & \cdot & -\sin\theta\cos\theta \\ \cdot & \frac{1}{r} & \cot\theta & \cdot \end{bmatrix}$$

Fig. 8.1

A'', etc., for the derivatives with respect to r.)

$$\Gamma^t{}_{tr} = \Gamma^t{}_{rt} = \frac{A'}{2A}$$

$$\Gamma^r{}_{tt} = \frac{A'}{2B} \qquad \Gamma^r{}_{rr} = \frac{B'}{2B} \qquad \Gamma^r{}_{\theta\theta} = -\frac{r}{B} \qquad \Gamma^r{}_{\phi\phi} = -\frac{r\sin^2\theta}{B}$$

$$\Gamma^\theta{}_{r\theta} = \Gamma^\theta{}_{\theta r} = \frac{1}{r} \qquad \Gamma^\theta{}_{\phi\phi} = -\sin\theta\cos\theta$$

$$\Gamma^\phi{}_{r\phi} = \Gamma^\phi{}_{\phi r} = \frac{1}{r} \qquad \Gamma^\phi{}_{\theta\phi} = \Gamma^\phi{}_{\phi\theta} = \cot\theta. \tag{8.9}$$

The other 51 Γs are all zero.

Using these results, we may assemble the matrices of Fig. 8.1 in a way that should be familiar by now.

8.6 Calculating the Riemann tensor

There are six **B**-matrices to be calculated according to (6.14). Perhaps the most tedious is

$$\mathbf{B}_{tr} = \partial_t\mathbf{\Gamma}_r - \partial_r\mathbf{\Gamma}_t + \mathbf{\Gamma}_t\mathbf{\Gamma}_r - \mathbf{\Gamma}_r\mathbf{\Gamma}_t.$$

The four individual terms are

$$\partial_t\mathbf{\Gamma}_r = 0, \quad -\partial_r\mathbf{\Gamma}_t = \begin{bmatrix} \cdot & -A''/2A + A'^2/2A^2 & \cdot & \cdot \\ -A''/2B + A'B'/2B^2 & \cdot & \cdot & \cdot \\ \cdot & \cdot & \cdot & \cdot \\ \cdot & \cdot & \cdot & \cdot \end{bmatrix},$$

$$\mathbf{\Gamma}_t\mathbf{\Gamma}_r = \begin{bmatrix} \cdot & A'B'/4AB & \cdot & \cdot \\ A'^2/4AB & \cdot & \cdot & \cdot \\ \cdot & \cdot & \cdot & \cdot \\ \cdot & \cdot & \cdot & \cdot \end{bmatrix}, \quad -\mathbf{\Gamma}_r\mathbf{\Gamma}_t = \begin{bmatrix} \cdot & -A'^2/4A^2 & \cdot & \cdot \\ -A'B'/4B^2 & \cdot & \cdot & \cdot \\ \cdot & \cdot & \cdot & \cdot \\ \cdot & \cdot & \cdot & \cdot \end{bmatrix}.$$

Summing these contributions gives

$$\mathbf{B}_{tr} = \begin{bmatrix} \cdot & -A''/2A + A'^2/4A^2 + A'B'/4AB & \cdot & \cdot \\ -A''/2B + A'^2/4AB + A'B'/4B^2 & \cdot & \cdot & \cdot \\ \cdot & \cdot & \cdot & \cdot \\ \cdot & \cdot & \cdot & \cdot \end{bmatrix}.$$

Additionally, the inverse of the metric matrix is

$$\mathbf{g}^{-1} = \begin{bmatrix} 1/A & \cdot & \cdot & \cdot \\ \cdot & -1/B & \cdot & \cdot \\ \cdot & \cdot & -1/r^2 & \cdot \\ \cdot & \cdot & \cdot & -1/r^2 \sin^2\theta \end{bmatrix},$$

so that

$$\mathbf{C}_{tr} = \mathbf{B}_{tr}\mathbf{g}^{-1}$$

$$= \begin{bmatrix} \cdot & A''/2AB - A'^2/4A^2B - A'B'/4AB^2 & \cdot & \cdot \\ -A''/2AB + A'^2/4A^2B + A'B'/4AB^2 & \cdot & \cdot & \cdot \\ \cdot & \cdot & \cdot & \cdot \\ \cdot & \cdot & \cdot & \cdot \end{bmatrix}$$

which is antisymmetric, as required. We may immediately pick out the nonzero component of the Riemann tensor, in row t and column r,

$$R^{tr}{}_{tr} = \frac{A''}{2AB} - \frac{A'B'}{4AB^2} - \frac{A'^2}{4A^2B}.$$

The other matrices are obtained similarly, and are listed separately in Fig. 8.2.

$$\begin{bmatrix} [C_{tt}] & [C_{tr}] & [C_{t\theta}] & [C_{t\varphi}] \\ [C_{rt}] & [C_{rr}] & [C_{r\theta}] & [C_{r\varphi}] \\ [C_{\theta t}] & [C_{\theta r}] & [C_{\theta\theta}] & [C_{\theta\varphi}] \\ [C_{\varphi t}] & [C_{\varphi r}] & [C_{\varphi\theta}] & [C_{\varphi\varphi}] \end{bmatrix} = \left[\begin{matrix}
\begin{bmatrix} \cdot&\cdot&\cdot&\cdot \\ \cdot&\cdot&\cdot&\cdot \\ \cdot&\cdot&\cdot&\cdot \\ \cdot&\cdot&\cdot&\cdot \end{bmatrix} &
\begin{bmatrix} \cdot&a&\cdot&\cdot \\ -a&\cdot&\cdot&\cdot \\ \cdot&\cdot&\cdot&\cdot \\ \cdot&\cdot&\cdot&\cdot \end{bmatrix} &
\begin{bmatrix} \cdot&\cdot&b&\cdot \\ \cdot&\cdot&\cdot&\cdot \\ -b&\cdot&\cdot&\cdot \\ \cdot&\cdot&\cdot&\cdot \end{bmatrix} &
\begin{bmatrix} \cdot&\cdot&\cdot&b \\ \cdot&\cdot&\cdot&\cdot \\ \cdot&\cdot&\cdot&\cdot \\ -b&\cdot&\cdot&\cdot \end{bmatrix} \\
\begin{bmatrix} \cdot&-a&\cdot&\cdot \\ a&\cdot&\cdot&\cdot \\ \cdot&\cdot&\cdot&\cdot \\ \cdot&\cdot&\cdot&\cdot \end{bmatrix} &
\begin{bmatrix} \cdot&\cdot&\cdot&\cdot \\ \cdot&\cdot&\cdot&\cdot \\ \cdot&\cdot&\cdot&\cdot \\ \cdot&\cdot&\cdot&\cdot \end{bmatrix} &
\begin{bmatrix} \cdot&\cdot&\cdot&\cdot \\ \cdot&\cdot&c&\cdot \\ \cdot&-c&\cdot&\cdot \\ \cdot&\cdot&\cdot&\cdot \end{bmatrix} &
\begin{bmatrix} \cdot&\cdot&\cdot&\cdot \\ \cdot&\cdot&\cdot&c \\ \cdot&\cdot&\cdot&\cdot \\ \cdot&-c&\cdot&\cdot \end{bmatrix} \\
\begin{bmatrix} \cdot&\cdot&-b&\cdot \\ \cdot&\cdot&\cdot&\cdot \\ b&\cdot&\cdot&\cdot \\ \cdot&\cdot&\cdot&\cdot \end{bmatrix} &
\begin{bmatrix} \cdot&\cdot&\cdot&\cdot \\ \cdot&\cdot&-c&\cdot \\ \cdot&c&\cdot&\cdot \\ \cdot&\cdot&\cdot&\cdot \end{bmatrix} &
\begin{bmatrix} \cdot&\cdot&\cdot&\cdot \\ \cdot&\cdot&\cdot&\cdot \\ \cdot&\cdot&\cdot&\cdot \\ \cdot&\cdot&\cdot&\cdot \end{bmatrix} &
\begin{bmatrix} \cdot&\cdot&\cdot&\cdot \\ \cdot&\cdot&\cdot&\cdot \\ \cdot&\cdot&\cdot&d \\ \cdot&\cdot&-d&\cdot \end{bmatrix} \\
\begin{bmatrix} \cdot&\cdot&\cdot&-b \\ \cdot&\cdot&\cdot&\cdot \\ \cdot&\cdot&\cdot&\cdot \\ b&\cdot&\cdot&\cdot \end{bmatrix} &
\begin{bmatrix} \cdot&\cdot&\cdot&\cdot \\ \cdot&\cdot&\cdot&-c \\ \cdot&\cdot&\cdot&\cdot \\ \cdot&c&\cdot&\cdot \end{bmatrix} &
\begin{bmatrix} \cdot&\cdot&\cdot&\cdot \\ \cdot&\cdot&\cdot&\cdot \\ \cdot&\cdot&\cdot&-d \\ \cdot&\cdot&d&\cdot \end{bmatrix} &
\begin{bmatrix} \cdot&\cdot&\cdot&\cdot \\ \cdot&\cdot&\cdot&\cdot \\ \cdot&\cdot&\cdot&\cdot \\ \cdot&\cdot&\cdot&\cdot \end{bmatrix}
\end{matrix}\right]$$

$a = A''/2AB - A'B'/4AB^2 - A'^2/4A^2B$
$b = A'/2rAB$
$c = -B'/2rB^2$
$d = (B-1)/r^2B$

A dot denotes zero.

$[C_{\varrho\sigma}]^{\alpha\beta} \equiv R^{\alpha\beta}{}_{\varrho\sigma}$

Fig. 8.2

In summary, the Riemann tensor has 24 nonzero components, all of which are visible in Fig. 8.2. Six are listed here:

$$R^{tr}{}_{tr} = \frac{A''}{2AB} - \frac{A'B'}{4AB^2} - \frac{A'^2}{4A^2B} \qquad R^{r\theta}{}_{r\theta} = -\frac{B'}{2rB^2}$$

$$R^{t\theta}{}_{t\theta} = \frac{A'}{2rAB} \qquad R^{r\phi}{}_{r\phi} = -\frac{B'}{2rB^2}$$

$$R^{t\phi}{}_{t\phi} = \frac{A'}{2rAB} \qquad R^{\theta\phi}{}_{\theta\phi} = -\frac{B-1}{r^2B}. \tag{8.10}$$

The remaining 18 are immediately obtained via the antisymmetry properties of the Riemann tensor. The six values happen also to be precisely the principal curvatures; the fact that certain pairs of values are equal reflects the spherical symmetry, as does the fact that they are all independent of θ and ϕ.

8.7 The vacuum field equations and the Schwarzschild solution

Only four of the components of the Ricci tensor are nonzero; they are

$$R^t{}_t = \frac{A''}{2AB} - \frac{A'B'}{4AB^2} - \frac{A'^2}{4A^2B} + 2\frac{A'}{2rAB},$$

$$R^r{}_r = \frac{A''}{2AB} - \frac{A'B'}{4AB^2} - \frac{A'^2}{4A^2B} - 2\frac{B'}{2rB^2},$$

$$R^\theta{}_\theta = R^\phi{}_\phi = \frac{A'}{2rAB} - \frac{B'}{2rB^2} - \frac{B-1}{r^2B}. \tag{8.11}$$

Thus the Einstein equations Ricci $= 0$ give three equations for the two unknown functions A and B. (In fact, the equations are not independent, by virtue of the obligatory Bianchi identities.)

We choose to solve a certain combination in which A and its derivatives are absent:

$$R^t{}_t - R^r{}_r - R^\theta{}_\theta - R^\phi{}_\phi = 2\frac{B'}{rB^2} + 2\frac{B-1}{r^2B} = 0.$$

Rearranging

$$\frac{\mathrm{d}B}{B(B-1)} + \frac{\mathrm{d}r}{r} = 0.$$

Integrating yields a result with one disposable constant a,

$$B = (1 - a/r)^{-1}.$$

To find A, we employ the combination

$$R^t{}_t - R^r{}_r = \frac{A'}{rAB} + \frac{B'}{rB^2} = 0,$$

implying

$$\frac{\mathrm{d}A}{A} = \frac{\mathrm{d}B}{B} = 0.$$

Integration gives immediately,

$$AB = \text{constant}.$$

As $r \to \infty$, clearly $B \to 1$ in any case. The requirement that $A \to c^2$ now leads to

$$A = c^2(1 - \ a/r).$$

This completes the solution, and we finally arrive at the familiar spherically symmetric, static, asymptotically-Minkowski solution of the Einstein vacuum equations: the Schwarzschild metric

$$c^2\,\mathrm{d}\tau^2 = c^2(1 - a/r)\,\mathrm{d}t^2 - \frac{\mathrm{d}r^2}{1 - a/r} - r^2(\mathrm{d}\theta^2 + \sin^2\theta\,\mathrm{d}\phi^2),$$

with just one disposable parameter a.

Substituting the functions A and B in (8.10) gives six elements of the Riemann tensor

$$R^{tr}{}_{tr} = R^{\theta\phi}{}_{\theta\phi} = -\frac{a}{r^3}, \quad R^{t\theta}{}_{t\theta} = R^{t\phi}{}_{t\phi} = R^{r\theta}{}_{r\theta} = R^{r\phi}{}_{r\phi} = \frac{a}{2r^3}. \tag{8.12}$$

8.8 The tidal tensor: an example

An artificial satellite is in free radial fall towards the Earth; what internal stresses result from tidal forces?

For simplicity, we shall imagine that the satellite is momentarily at rest at the radial position r. The contravariant velocity is then

$$\dot{t} = \text{constant}, \quad \dot{r} = 0, \quad \dot{\theta} = 0, \quad \dot{\phi} = 0.$$

One of the sixteen elements of $\boldsymbol{\Delta}$—which happens to be nonzero—is

$$\Delta^r{}_r = \dot{t}\dot{t}R^r{}_{ttr} + \dot{t}\dot{r}R^r{}_{trr} + \cdots + \dot{\phi}\dot{\phi}R^r{}_{\phi\phi r} \quad \text{(sixteen terms in all).}$$

Only the first term survives; here

$$R^r{}_{ttr} = g_{tt}R^{rt}{}_{tr} = -c^2\left(1 - \frac{a}{r}\right)R^{tr}{}_{tr}.$$

(Note how an antisymmetry in R yields the minus sign.) Since the velocity is to be a 'unit' vector, $c^2(1 - a/r)\dot{t}^2 = c^2$, and finally

$$\Delta^r{}_r = \frac{ac^2}{r^3}.$$

The remaining elements may be found similarly. Thirteen are zero, and the other two are

$$\Delta^\theta{}_\theta = \Delta^\phi{}_\phi = -\frac{ac^2}{2r^3}.$$

The signs show that the satellite is under radial tension and transverse compression, in complete accord with the expectations of section 1.13. In Earth orbit, such forces are easily measurable with suitable instrumentation, and are valuable for purposes of practical navigation (Fig. 8.3).

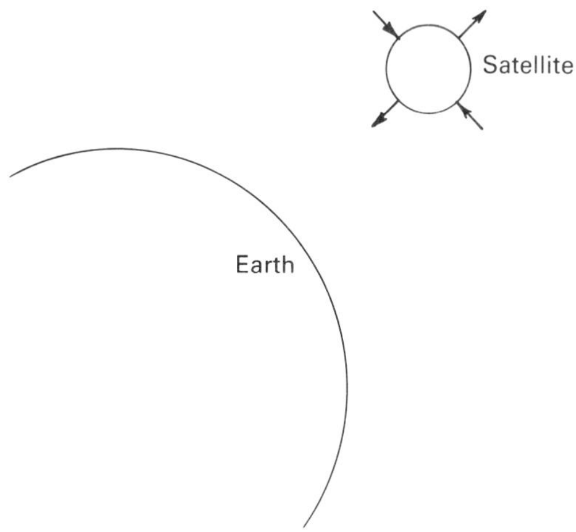

Fig. 8.3

The assumption of momentary rest is unnecessary; the stresses for any radial free fall are the same.

8—Notes and problems

1. Show that $\Delta_{\mu\nu}$ and $R_{\mu\nu}$ are both symmetric. (Both results depend on $R_{\alpha\beta\gamma\delta} = R_{\gamma\delta\alpha\beta}$.) Show that $\Delta_{\mu\nu}$ is *spacelike*, that is, $\Delta_{\mu\nu}\dot{x}^\nu = 0$. (This is required if the tidal force is to be a spacelike vector.)
2. Complete the work of sections 8.5 and 8.6 to obtain Figs. 8.1 and 8.2.
3. **The 'Schwarzschild' solution with the cosmological term** Show that the inclusion of the cosmological term (equation (8.5)) in the search for a spherically symmetric metric leads to the requirements

$$\frac{B'}{rB^2} + \frac{B-1}{r^2B} = \Lambda \quad \text{and} \quad AB = \text{constant}.$$

(Use the same combinations as in section 8.7.)

We are already well on the way to the solution of the first of these, since we know the complete answer when Λ is zero. Substitute $B = 1/(1 - a(r)/r)$ to obtain the very simple differential equation for $a(r)$; hence find

$$B = 1/\{1 - a/r - \Lambda r^2/3\}, \quad \text{with } a = \text{constant again.}$$

Verify that the six principal curvatures are now $a/2r^3 - \Lambda/3$ (four times) and $-a/r^3 - \Lambda/3$ (twice).

9

Black holes

9.1 What about $r = a$?

Already we have made some use of the Schwarzschild metric as a description of the spacetime in the Solar System, and we have just shown how it follows from the Einstein field equations for empty space. Inspection immediately shows that there is some kind of trouble when $r = 0$ and $r = a$, since elements of the metric misbehave at these values. We must not jump to the conclusion that there need be anything wrong with spacetime itself; misbehaviour in the metric may just as well arise from an unsatisfactory choice of coordinates. How can we tell?

It is elementary to verify that the six principal curvatures take the values of (8.12)

$$-\frac{a}{r^3} \text{ (twice)}, \quad \frac{a}{2r^3} \text{ (four times)}.$$

Clearly, each of these is singular at $r = 0$. Clearly therefore *Schwarzschild spacetime must be singular at* $r = 0$.

Remarkably, none of these curvatures is singular at $r = a$, in spite of the awkwardness in the coefficients in the metric. We are led to the suggestion that there may be no singularity in the spacetime itself at that radius; all that has happened is that we have chosen coordinates which are not able to represent the spacetime adequately at $r = a$.

The first step towards a solution for this problem was taken by Eddington and Finkelstein in the early 1930s, who removed the coordinate singularities in a partial and rather unsymmetric way. Strangely, it was not till 1961 that a full resolution became available; it is surprisingly simple.

9.2 The worldlines of radially moving photons

One technique for circumventing the problems of unsatisfactory coordinates is to probe the space with geodesics which, after all, are coordinate-independent and will not be affected in any way by the boundaries of coordinate validity. Out of many possibilities we shall use as probes the null-worldlines of radially moving photons.

We begin in the Schwarzschild coordinate system, which is the only one we know. On any null radial geodesic ($d\theta$ and $d\phi$ both zero) we have

$$0 = c^2(1 - a/r)\,dt^2 - \frac{dr^2}{1 - a/r}$$

whence either

$$c\,dt = \frac{dr}{1 - a/r} \quad \text{for a } \textit{rising} \text{ photon,}$$

or

$$-c\,dt = \frac{dr}{1 - a/r} \quad \text{for a } \textit{falling} \text{ photon.}$$

These immediately integrate to

$$\begin{aligned} ct &= r + a\ln(r - a) + C \quad \text{(rising photon)} \\ -ct &= r + a\ln(r - a) + D \quad \text{(falling photon).} \end{aligned} \tag{9.1}$$

where C and D are constants of integration.

The trick is to use the constants of integration as new coordinates. Since $C = \text{const}$ along the entire worldline of a rising photon, C will be a 'good' coordinate wherever that worldline penetrates; the same is true for D and a falling photon. Indeed, we can do rather better by using the closely related coordinates U and V defined by

$$UV = e^{r/a}\left(\frac{r}{a} - 1\right), \quad \frac{V}{U} = e^{ct/a}. \tag{9.2}$$

(these are nearly the coordinates introduced by Kruskal and Szekeres in 1961). In the familiar region $r > a$, both U and V must be *positive*, but there is no reason to prohibit a passage to negative values. In this way we discover a more complete description of the Schwarzschild space.

For convenience, plot U and V as Cartesian coordinates as shown in Fig. 9.1. (The plot is of those events for which the polar angles θ and ϕ take **fixed** values.) The original coordinate grid now appears as a network of hyperbolas ($r = \text{constant} > a$) and straight lines ($t = \text{constant}$), confined to one quadrant of the diagram. On the U and V axes, $UV = 0$, and thus $r = a$: these axes represent the limits of validity of the old coordinates. On the other hand, the photon geodesics do not 'see' these axes; there the spacetime looks just as ordinary as it does elsewhere. What the geodesics *do* see is the singular line at $r = 0$ (that is, at $UV = -1$). This is a hyperbola whose future branch (the **future singularity**) the falling photon cannot avoid, and at whose past branch (the **past singularity**) all the rising photons originate. The diagram stops there; the singularities are genuine, and cannot be surmounted by any further adjustments.

Any attempt to extend the space with new geodesics leads nowhere, and the coordinates U and V (subject only to $UV > -1$) cover the whole of the Schwarzschild space.

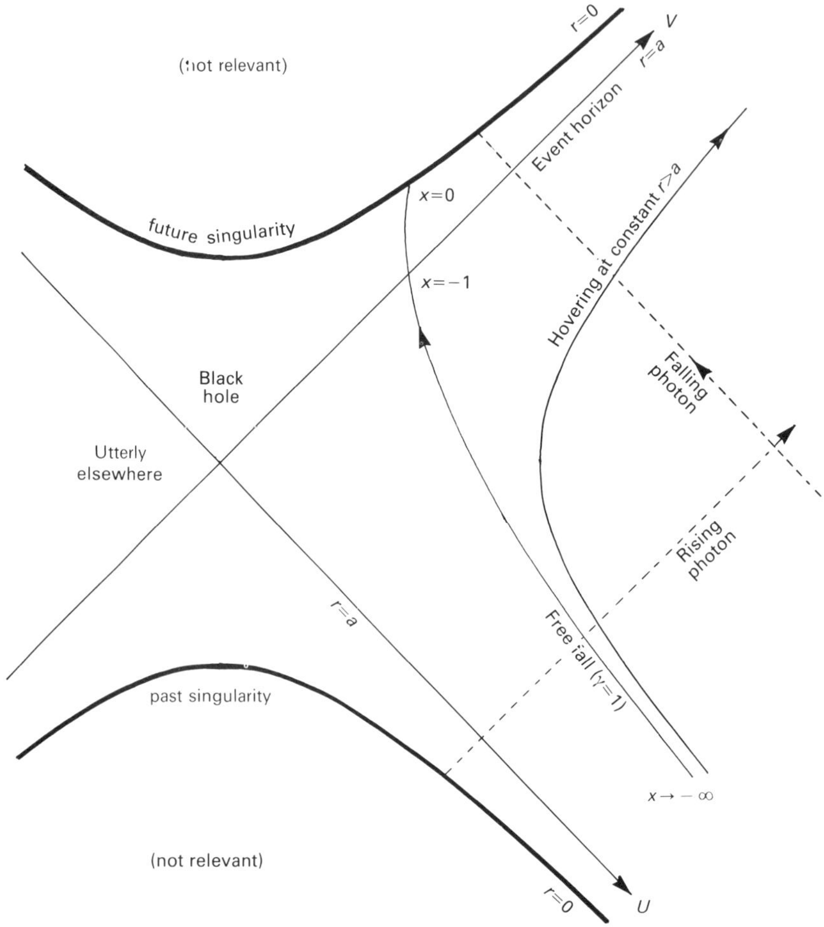

Fig. 9.1

9.3 The event horizon

Note the presence of the **event horizon**, the positive half of the V-axis. Because of the way in which Fig. 9.1 has been set out, all the lightcones have the same shape and the same orientation. It is thus very clear that any signal or other emission from an event lying to the future of the event horizon *must be trapped by the singularity*. There is no escape. The future lies in the direction of ever-decreasing r, and is closed off in nothingness. Hence the picturesque name **black hole**: everything goes in, nothing comes out—not even light to see by.

It is illuminating to ask what happens to an observer who is careless enough to get close to a black hole. The simplest case is radial free fall; the Lagrangian and a first integral have been given already in section 4.3. We shall consider the particularly straightforward case $\gamma = 1$, for which

$$\dot{r}^2 = \frac{ac^2}{r}, \tag{9.3}$$

and

$$\dot{t} = \frac{1}{1 - a/r}. \tag{9.4}$$

It is useful to introduce the (nonaffine) parameter x by

$$r = ax^2; \tag{9.5}$$

this leads to

$$2x^2\,\mathrm{d}x = c\,\mathrm{d}\tau$$

with solution

$$c\tau = 2ax^3/3.$$

(The sign and the arbitrary constant of integration have been chosen to obtain the desired solution.) As x goes from $-\infty$ to 0, the observer falls from a remote distance to the origin, while his proper time τ advances steadily, without interruption, from the remote past to zero. He will perceive no hint of any trouble at the event horizon $r = a$ (though disaster awaits him at $r = 0$, in the form of overwhelming tidal forces).

A second observer hovers at some distance from the black hole, and observes the history of the first observer, by radar, perhaps. In particular, he naturally measures progress in terms of the *coordinate* time t. From (9.4), $\mathrm{d}t = \mathrm{d}\tau/(1 - a/r)$, and substitution of r and τ in terms of x gives

$$c\,\mathrm{d}t = 2a\frac{x^4\,\mathrm{d}x}{x^2 - 1}.$$

Integrating,

$$ct = a\left(\frac{2}{3}x^3 + 2x + \ln\frac{x-1}{x+1}\right).$$

In the remote past ($x \to -\infty$), the dominant behaviours of t and τ are the same. However, as x increases towards -1 (where $c\tau = -2a/3$), $t \to +\infty$, and thereafter loses all meaning. Thus the hovering observer will never see the falling observer reach the horizon, since in his reckoning such an event lies in the remote future.

9.4 But is it physics?

Our Sun is closely spherically symmetric, and its vacuum environment is therefore very nearly Schwarzschild. Its nonvacuum interior is not. Later we shall discuss the nature of gravity in the inside of a nonzero mass density; for the moment it is enough to remark that there are no singularities and no event horizons inside the Sun. The best we can do just now is to draw a rather incomplete diagram (see Fig. 9.2).

However, one awesome possibility is **gravitational collapse**, in which a mass, larger than a certain critical value, cannot maintain enough pressure to hold up against its own gravitational self-attraction. This may happen with a large star as it cools towards the end of its life; it may begin to shrink to such an extent that its surface ends up inside the Schwarzschild radius $r = a$. Nothing can then prevent its collapse to the singularity. This process will finish in a finite proper time, though a distant

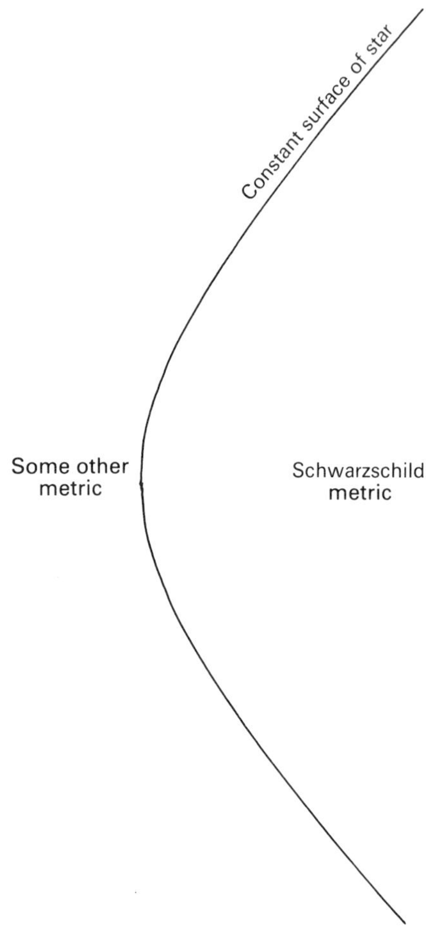

Fig. 9.2

observer will never be able to see its completion on account of the event horizon. The relevant diagram is a little more complete (see Fig. 9.3 and Chapter 12).

Can we envisage circumstances where the *whole* diagram is relevant (a black hole without matter)? Of course, we can: what we would have in mind is a **primordial black hole**, present from the start of the Universe. We shall need to remember, however, that much would have to be left to the imagination, since it is implied that there are parts of the Universe, of infinite extent, which will be for ever closed to us as observers: we can never see them, and we can never influence them. They are utterly elsewhere.

9.5 Photon orbits

We have used the null-worldlines of radially moving photons to explore the geometry of a spherical black hole. To form some idea of the *visual effects* of the presence of a black hole it is necessary to examine the null-worldlines of photons with more general motions.

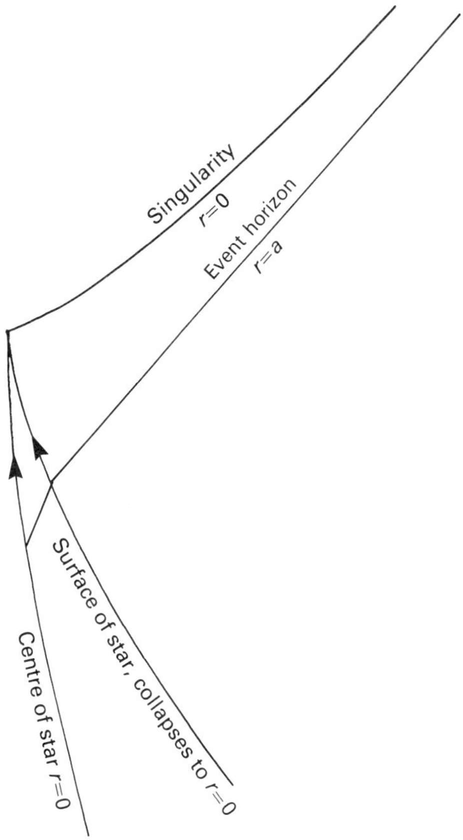

Fig. 9.3

It will be sufficient to consider the *orbits* of photons in the presence of the hole. In section 4.8, such orbits were handled using the coordinates r and ϕ. (As usual, the spherical symmetry of the system permits us to confine our thinking to equatorial orbits, for which $\theta = \pi/2$ throughout.) We are able to continue to use these coordinates, as r is a 'good' coordinate throughout the spacetime; in particular, it shows no undesirable behaviour at $r = a$. Thus we may adopt without change the exact equation

$$\left(\frac{du}{d\phi}\right)^2 = au^3 - u^2 + 1/b^2, \quad u \equiv 1/r, \quad b \equiv \text{impact parameter}, \tag{9.6}$$

obtained in section 4.8, and allow all positive values of r. However, it ought to be remembered that when $r < a$, r is a 'timelike' coordinate, and the relation between r and ϕ does not then describe the 'shape of the orbit' as normally understood.

The full solution involves elliptic functions, and it is no longer possible to rely on arguments which depend on the smallness of a/r. A qualitative picture of the shape of the orbits is obtained by examining the sign of the right side of (9.6): this right side must be nonnegative throughout any possible motion. In the rb-diagram (Fig. 9.4), the prohibited region is shown, and it is straightforward to identify the different types of orbit that can occur. (Examples for each case are shown in Fig. 9.5.) They are

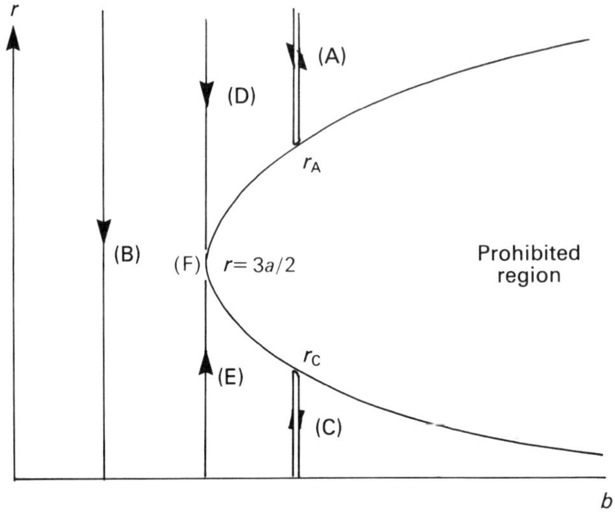

Fig. 9.4

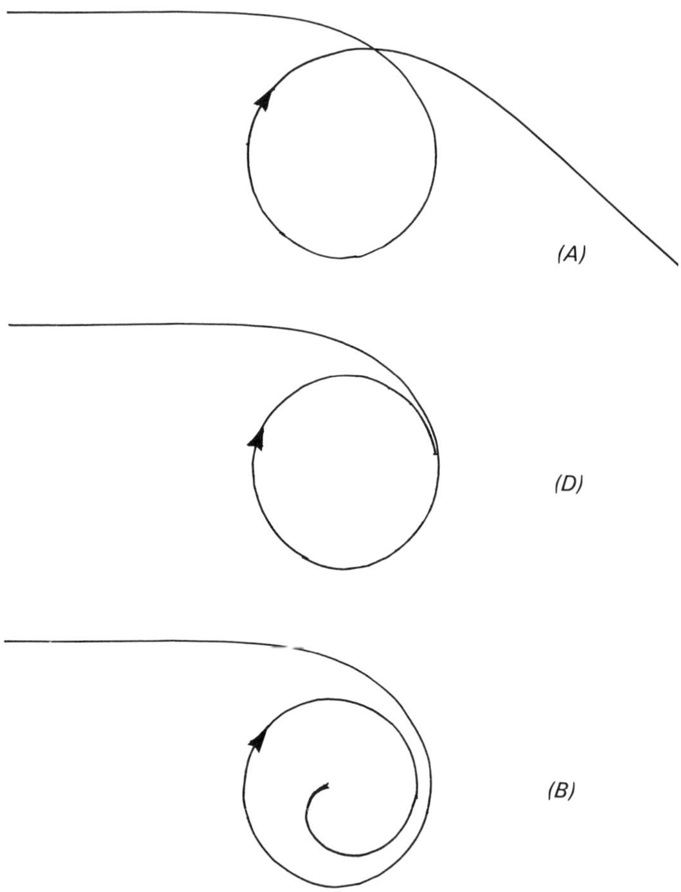

Fig. 9.5

(A) $b^2 > 27a^2/4$ For a large enough impact parameter b the photon arrives from infinity, reaches a pericentre r_A (always greater than $3a/2$) where the right side is zero, and returns to infinity. (This is in fact the case of the deflection of light by the Sun, dealt with in section 4.8. There the impact parameter is huge in comparison with a; the resulting smallness of a/r was used as the basis for a very good first-order approximation.)

(B) $b^2 < 27a^2/4$ If the impact parameter is small enough, the right side is positive for *all* values of r. The photon arrives from infinity, falls unhindered through the event horizon at $r = a$, and ends its life with utter finality at the future singularity $r = 0$. The reversed motion (B′) is also possible: the photon starts at the past singularity $r = 0$, rising steadily to infinity. (This case includes, for $b = 0$, the trajectories of the radially moving photons with which we explored the inner reaches of the Schwarzschild solution, section 9.2.)

(C) $b^2 > 27a^2/4$ (Note that this condition on b is the same as that in (A) above.) For such values of b a *second* orbit is always possible, in which the photon rises from the past singularity, reaches an apcentre r_C, and falls back to the future singularity. Always the apcentre satisfies $a < r_c < 3a/2$. For an orbit of this kind the significance of b as an impact parameter is evidently lost.

The critical case, where $b^2 = 27a^2/4$, is of particular interest. (Incidentally, it may be dealt with in terms of *elementary* functions alone.) There are several possible motions:

(D) The photon arrives from infinity (with impact parameter $b = \sqrt{(27/4)}a$), and spirals asymptotically down to the circle at $r = 3a/2$. It never reaches this circle, and never returns to infinity. The reversed motion (D′) is also possible.

(E) The photon rises from the past singularity, and spirals up to the circle at $r = 3a/2$, again never actually reaching it. In the reversed motion (E′), the photon starts near the circle, and spirals down to the future singularity.

(F) The limiting circle at $r = 3a/2$ is itself a possible orbit for a photon; at this radius, the gravity field of the black hole is strong enough to hold anything moving even at the speed of light. This orbit is unstable: the slightest of perturbations will be sufficient to change the orbit to one of the spirals (D′) or (E′).

9.6 How big is a black hole?

Black holes 'look' bigger than they 'are'. To be precise, think of a bundle of light rays, mutually parallel at infinity, travelling towards a black hole of 'size' a. Those with impact parameter greater than the critical value $a\sqrt{(27/4)}$ are restricted by a pericentre, and ultimately go off to infinity again. If the impact parameter is less than the critical value, the rays are not restricted, and descend all the way to $r = 0$. Reversing all the rays—as is evidently possible—shows that, when viewed from a large distance, the apparent radius of the hole is not a, nor even $3a/2$, but $a\sqrt{(27/4)}$.

This is **gravitational lensing**, in which, as a result of their gravity field, objects appear to be larger than they are. The effect of this lensing on the *background* of the object is complicated by the fact that a ray whose perihelion is slightly greater than

$3a/2$ may circumnavigate the hole many times before escaping again to infinity; consequently the image of the hole itself is fringed with a confusion of multiple images of the rest of the Universe. Further out, the confusion is replaced by a simple magnification of the background: it was this that was measured by Eddington in his crucial observations of the solar eclipses of 1919.

The problem may be generalized somewhat to yield the angular size of a black hole when viewed by an observer A hovering at a constant Schwarzschild 'distance' r. An orbit passing through A makes an angle χ with the radial direction; a simple geometrical relation is

$$\tan\chi = \frac{r\,\mathrm{d}\phi}{\mathrm{d}r\sqrt{(1-a/r)}}.$$

A simple rearrangement gives

$$\cos\chi = \frac{\mathrm{d}u/\mathrm{d}\phi}{\sqrt{[(\mathrm{d}u/\mathrm{d}\phi)^2 + u^2 - au^3]}}$$

(this is a general result for all free fall orbits, whether null or not). The angular radius of the black hole is obtained by replacing $\mathrm{d}u/\mathrm{d}\phi$ by the expression appropriate to the spiral photon orbit,

$$\begin{aligned}(\mathrm{d}u/\mathrm{d}\phi)^2 &= au^3 - u^2 + 4/27a^2\\ &\equiv a(u-2/3a)^2(u+1/3a),\end{aligned}$$

giving, after a slight rearrangement,

$$\cos\chi = (1 - 3a/2r)\sqrt{(1+3a/r)}.$$

As expected, as $r\to\infty$, $\chi\to 0$: the hole shrinks in angular size. Now move to the much closer position $r = 3a/2$; χ increases to $\pi/2$, and the black hole fills half of the sky. As the observer descends to the event horizon, χ continues to increase, and the hole appears to wrap itself threateningly round the observer, leaving an ever-decreasing cone of sight to the Universe 'outside'. At the horizon, the outside can no longer be seen.

All this is for the hovering observer. An observer in free radial fall sees things differently; he is still aware of the Universe outside even after crossing the event horizon, and his cone of sight—though diminishing—never tends to nothing, even when the singularity at $r = 0$ is reached.

Reversing the orbits gives information about the behaviour of a stationary *emitter* in the neighbourhood of a black hole. All radiation emitted within the cone of semiangle χ is captured by the hole, and the nearer the emitter is to $r = a$, the smaller the proportion of the emission that finds its way to infinity. Thus it is to be expected that the emitter will become very faint when near the event horizon. This effect is in addition to the gravitational redshift whereby the emitted photons arrive at infinity with much reduced energy anyway.

9.7 Orbits of massive particles

We shall look briefly at the more general case of timelike worldlines, whose orbits require two disposable constants.

The relevant equation is (4.9):

$$\left(\frac{\mathrm{d}u}{\mathrm{d}\phi}\right)^2 = au^3 - u^2 + \alpha u + \beta. \tag{9.7}$$

As we know, the full solution involves elliptic functions. We shall be content with a qualitative discussion.

Equations (4.10) show that

$$\alpha \geqslant 0, \quad \alpha + a\beta \geqslant 0;$$

there are no other restrictions on α and β. The right side of (9.7) must be nonnegative throughout the motion, and thus the qualitative features of the orbits depend crucially on the positions of the real positive roots of the cubic

$$au^3 - u^2 + \alpha u + \beta = 0 \tag{9.8}$$

(such roots prescribe any apcentres or pericentres there may be). It is therefore useful to plot the $(\alpha\beta)$-locus for *equal* roots, separating the region for three real roots from the region for only one (Fig. 9.6). The equation for this locus is

$$27(\beta a^2)^2 + 18(\alpha a)(\beta a^2) - 4\beta a^2 + 4(\alpha a)^3 - (\alpha a)^2 = 0. \tag{9.9}$$

The line $\beta = 0$ is also important, as separating the orbits which extend to infinity ($u = 0$) from those which do not; it is the locus across which a root changes sign.

The β-axis ($\alpha = 0$) relates to the photon orbits considered in section 9.5.

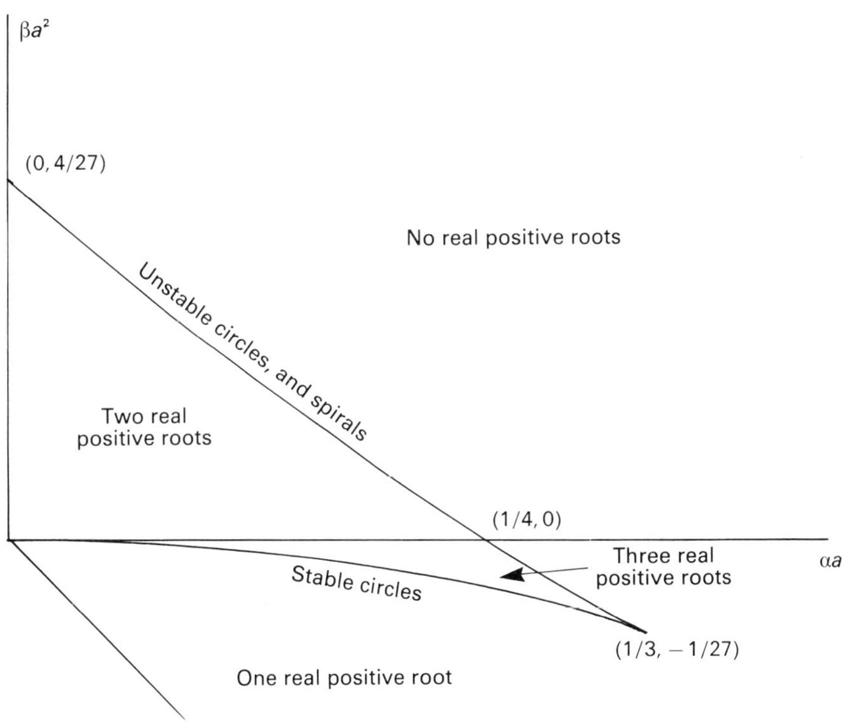

Fig. 9.6

The diagram is divided into four regions by the loci; in the different regions the orbits are qualitatively different:

No real positive roots. This implies that there are no pericentres and no apcentres. Any orbit associated with this region must extend all the way from $r=0$ to $r=\infty$.

One real positive root. There is just one restriction on the travel of an orbit, which turns out to be an **apcentre** r_0. Such an orbit begins at $r=0$ (at the past singularity, of course), rises to $r=r_0$, and then falls back to $r=0$ at the future singularity.

Two real positive roots. Here there is an apcentre r_0 and a pericentre r_1, with $r_0<r_1$. There are *two* possible orbits: one is as just described; the other comes from $r=\infty$ to the pericentre, before returning to ∞. (The 'hyperbolic' orbits of astronomy belong to this class.)

Three real positive roots. Again there is an apcentre r_0 which results in an internal orbit as before. The other roots comprise a pericentre r_1 and an apcentre r_2, which delimit an 'elliptic' orbit. (Planetary orbits belong here.)

In the photon case, the number of real positive roots had to be either zero or two. Now all possibilities occur, and a greater variety of orbits is found.

9.8 Circular orbits

Circular orbits are possible whenever two of the roots of the cubic coincide. Such a case corresponds to a point on the boundary between the regions. The orbit may be stable or unstable; when it is unstable, there exist further *spiral* orbits which approach the circle in an asymptotic manner. The circle itself may be regarded as an apcentre or pericentre which is never actually reached.

For a circular orbit, u is constant at all times. In particular, both $\mathrm{d}u/\mathrm{d}\phi$ and $\mathrm{d}^2u/\mathrm{d}\phi^2$ are zero; the first is given at (9.7), and differentiating gives immediately

$$\frac{\mathrm{d}^2u}{\mathrm{d}\phi^2}=3au^2-2u+\alpha.$$

Setting these to zero and solving gives

$$\alpha=2u-3au^2 \quad\text{and}\quad \beta=-u^2+2au^3; \tag{9.10}$$

clearly these constants, and therefore the complete motion, are fixed by the orbit radius $1/u$, which is now the only disposable parameter. Use of (4.10) yields straightforwardly

$$\dot{t}(1-au)=\gamma=\frac{1-au}{\sqrt{(1-3au/2)}} \quad\text{and}\quad \dot{\phi}/u^2=J=\sqrt{\left(\frac{ac^2}{2u-3au^2}\right)}$$

for the values of the first integrals of the motion in terms of u. Thus the contravariant fourvelocity components are

$$\dot{t}=\frac{1}{\sqrt{(1-3au/2)}},\quad \dot{r}=0,\quad \dot{\theta}=0,\quad \dot{\phi}=\frac{1}{\sqrt{(1-3au/2)}}\sqrt{\left(\frac{ac^2u^3}{2}\right)}; \tag{9.11}$$

all are constant, since u is. They are ill-defined for the circular photon orbit at $r=3a/2$, and meaningless at smaller radii: there can be no circular orbit smaller than the photon orbit.

For radii greater than $3a$, the circular orbits are stable, and slight perturbations have little effect. For an orbit whose radius lies between $3a/2$ and $3a$, any disturbance will send the particle off on a spiral, either winding down to $r = 0$, or climbing to an apcentre or perhaps to infinity. On account of the special nature of the case, the spiral orbits may be described in terms of 'elementary' functions alone.

9.9 Tidal stresses on a satellite in circular orbit

Knowing the four-velocity (9.11) and the Riemann tensor (8.12) allows us to calculate the tidal tensor Δ, according to

$$\Delta^{\mu}{}_{\nu} \equiv \dot{x}^{\rho}\dot{x}^{\sigma}R^{\mu}{}_{\rho\sigma\nu} = \dot{t}\dot{t}R^{\mu}{}_{tt\nu} + \dot{t}\dot{\phi}(R^{\mu}{}_{t\phi\nu} + R^{\mu}{}_{\phi t\nu}) + \dot{\phi}\dot{\phi}R^{\mu}{}_{\phi\phi\nu},$$

in which μ and ν each range over t, r, θ, ϕ; the other twelve terms on the right are absent. There are sixteen elements to be calculated; ten of them are zero. A typical calculation runs—omitting zero terms—

$$\begin{aligned}
\Delta^{r}{}_{r} &= -\dot{t}^{2}R^{rt}{}_{rt}g_{tt} - \dot{\phi}^{2}R^{r\phi}{}_{r\phi}g_{\phi\phi} \\
&= -\frac{1}{1-3au/2}\left(-\frac{a}{r^{3}}\right)c^{2}(1-au) - \frac{ac^{2}u^{3}}{2-3au}\left(\frac{a}{2r^{3}}\right)(-r^{2}) \\
&= \frac{ac^{2}}{r^{3}}\frac{2r-3a/2}{2r-3a};
\end{aligned}$$

the other elements may be obtained similarly. It is convenient to display the complete tensor as a matrix,

$$\Delta = \frac{ac^{2}}{r^{3}(2r-3a)}\begin{bmatrix} a/2 & \cdot & \cdot & -(r/c)\sqrt{(a/2r)} \\ \cdot & 2r-3a/2 & \cdot & \cdot \\ \cdot & \cdot & -r & \cdot \\ c(r-a)\sqrt{(a/2r)} & \cdot & \cdot & -r+a \end{bmatrix}.$$

As usual, the physical magnitude of the stress is best found from the eigenvalue relation

$$\Delta^{\mu}{}_{\nu}\lambda^{\nu} = \kappa\lambda^{\mu}.$$

One of the possible values of κ is zero (for $\lambda = \dot{x}$); the other three are the **principal stresses**, and are

$$\frac{ac^{2}}{r^{3}}\frac{2r-3a/2}{2r-3a}, \quad -\frac{ac^{2}}{r^{3}}\frac{r}{2r-3a}, \quad -\frac{ac^{2}}{2r^{3}}.$$

The fact that these are larger in magnitude (section 8.8) than those for a hovering object (and indeed become infinite for the photon orbit) may be put down to a centrifugal effect, though it makes little sense in the present context to try to separate centrifugal and gravity forces.

9—Notes and problems

1. The partial improvement of Eddington and Finkelstein retains the coordinates r, θ, and ϕ, but replaces t by a new coordinate

$$w = ct + r + a \ln(r/a - 1).$$

Comparison with (9.1) shows that the null-worldline of a radially falling photon is given by $w = \text{const}$. Sketch the coordinate grid, and show that just half of the black hole space is covered.

Show that the Eddington–Finkelstein version of the metric is

$$c^2 \, d\tau^2 = (1 - a/r) \, dw^2 - 2 \, dw \, dr - r^2(d\theta^2 + \sin^2 \theta \, d\phi^2).$$

2. The photon orbits D, E, and F (section 9.5) all satisfy

$$\left(\frac{du}{d\phi}\right)^2 = au^3 - u^2 + \frac{4}{27a^2}.$$

With the help of the substitution $v^2 = au + 1/3$, show that the equation may be solved in terms of 'elementary' functions only, and obtain the solution

$$\frac{v-1}{v+1} = Ce^{\pm\phi},$$

in which C is a constant, and the sign of ϕ depends on the sense of the description of the orbit.

Show that the three cases are given by D: $[C < 0, 1/\sqrt{3} < v < 1]$, E: $[C > 0, 1 < v < \infty]$, and F: $[C = 0, v = 1]$.

3. *All* massive objects look larger than they really are. Show that a light ray grazing the surface of a massive sphere of radius $R > 3a/2$ will arrive at infinity with impact parameter

$$b = R\sqrt{\left(\frac{R}{R-a}\right)}.$$

Show that the apparent diameter of the Sun exceeds the true diameter by nearly 3 km.

4. If $au^3 - u^2 + \alpha u + \beta = 0$ has a double root u, then its derivative $3au^2 - 2u + \alpha = 0$ must possess the same root. The condition that this should happen is found by eliminating u from the two equations. [You may prefer to start from the slightly rearranged (9.10).]

5. A spiral associated with the circular orbit of radius $1/u_0$ is governed by the equation

$$\left(\frac{du}{d\phi}\right)^2 = a(u - u_0)^2\left(u + 2u_0 - \frac{1}{a}\right).$$

Make the substitutions $v^2 = au + 2au_0 - 1$, ${v_0}^2 = 3au_0 - 1$, to obtain

$$2\frac{dv}{d\phi} = \pm({v}^2 - {v_0}^2)$$

with the general solution

$$\frac{v - v_0}{v + v_0} = Ce^{\pm v_0 \phi}.$$

Discuss the different cases on the lines of Problem 2.

When the radius of the circular orbit exceeds $3a$, there are no spirals. Nevertheless, with real v and pure imaginary v_0 and C, the general solution gives another orbit. Investigate it.

6. When v_0 is zero (radius $= 3a$) the general solution of the previous problem fails by becoming trivial. Show that the spiral orbit is now described by

$$u = \frac{1}{3a} + \frac{4}{a\phi^2}.$$

7. In the integral

$$\int_{\text{one orbit}} \frac{\mathrm{d}u}{\sqrt{\{(u - u_1)(u_2 - u)(1 - au_1 - au_2 - au)\}}},$$

make the substitution $u = (u_1 + u_2 q^2)/(1 + q^2)$ to obtain

$$2\pi + \varepsilon = \int_{-\infty}^{\infty} \frac{2\mathrm{d}q}{(1 + q^2)\sqrt{\left[1 - au_1 - au_2 - a\dfrac{u_1 + u_2 q^2}{1 + q^2}\right]}},$$

where ε is the perihelion precession per orbit, for a bound particle of nonzero mass.

Show that for a circular orbit of radius $1/u$ this integral is exactly $2\pi/\sqrt{(1 - 3au)}$. (Of course, a *circular* orbit cannot be said to exhibit a precession; interpret the result appropriately.)

Note that precession is meaningless for a circular orbit of radius less than $3a$, the radius at and below which spiral orbits are possible.

8. Two compact masses, each of 1 kg, are joined by a strong wire of length 2 m. The arrangement is placed 1 km away from a black hole with $a = 1$ cm, and allowed to fall; the masses are in line with the black hole, and the wire is taut. Show that the tension in the wire is not far short of a million Newtons. [Hint: $(L/2)(ac^2/r^3)$; see section 8.8.]

10

The matter tensor

10.1 The source of gravity

So far we have considered the effects of gravity on matter (geodesics, orbits, timekeeping, and so on), and the law obeyed by gravity in a vacuum. We have now to deal with matter in its more active role as the *source* of gravity. Newtonian gravity introduces its source in a very simple way: the Laplace equation for the potential in a vacuum is replaced by the more general Poisson equation;

$$\nabla^2\phi = 0 \quad \text{becomes} \quad \nabla^2\phi = \text{constant}.\rho$$

where $\rho(\mathbf{r})$ is the mass density at $\mathbf{r}$.

Analogously, we shall expect that the Einstein vacuum equation will need to be generalized:

$$R_{\mu\nu} = 0 \quad \text{becomes} \quad R_{\mu\nu} = \text{something appropriate connected with matter.}$$

This chapter is concerned with the search for an appropriate right side, which we shall expect to be some kind of *density*.

10.2 Densities in Special Relativity

A 'density' is an 'amount per unit volume', and is defined in the manner of

$$\text{amount in the (small 3-dimensional) volume } \mathrm{d}V = \text{density} \times \mathrm{d}V.$$

This is perfectly in order in 3D space, but needs some modification to take account of a new feature in 4D spacetime.

The new feature is that a small 3D volume in a 4D space is no longer a scalar, and it is best described by a covariant vector as follows. Take a 3D volume V, and displace it through the vector $\mathbf{dx}$. The 4D hypervolume swept out in the process *is* a scalar, and depends linearly on $\mathbf{dx}$ and on V itself. We may write this relationship as

$$\text{hypervolume swept out} = \mathrm{d}x^{\mu} V_{\mu},$$

thus introducing the covariant vector V_{μ} to represent the volume V both as to magnitude and as to orientation.

The necessary change to the definition of a density is now clear:

$$\text{the amount of } A \text{ in the small volume } \mathrm{d}V = (\text{density of } A)^{\mu} \cdot \mathrm{d}V_{\mu}.$$

In this way the 3D density in 4D space acquires an extra contravariant affix.

10.3 Particle density

Imagine a collection of particle worldlines, not necessarily straight, and not necessarily parallel, distributed densely enough to be regarded as an effectively continuous distribution. In the space $t = 0$ there is a volume V bounded by a surface S (Fig. 10.1); we ask, how many worldlines encounter the volume V? Alternatively, how many worldlines does the surface S embrace?

The answer N may be obtained by integrating a suitable density through the volume. The answer is a matter of simple counting, and is therefore clearly a scalar; the density therefore will have to be a contravariant vector $n^{\mu}(xyz)$, defined throughout the volume, such that for a small volume $\mathrm{d}V_{\mu}$, the contribution to N is

$$\mathrm{d}N = n^{\mu} \mathrm{d}V_{\mu}.$$

Then finally

$$N = \int_V n^{\mu} \, \mathrm{d}V_{\mu}.$$

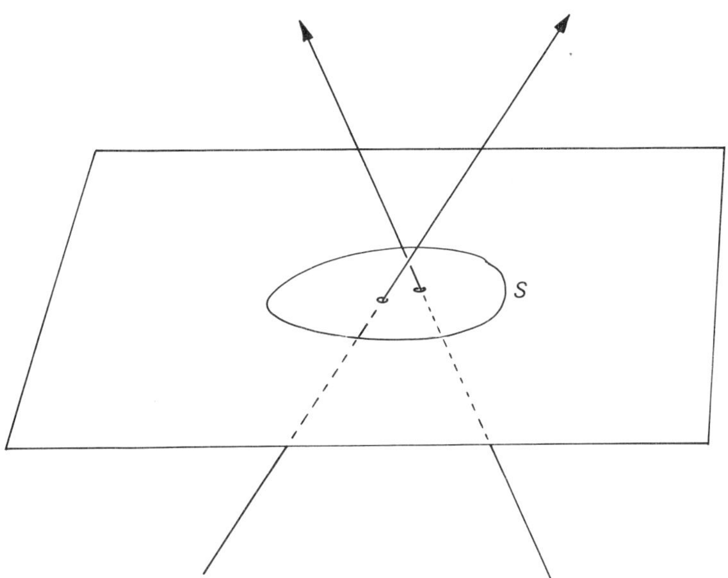

Fig. 10.1

Inasmuch as N is the number of intersections of particle worldlines with the volume V, it is exactly what we mean by a count of the number of particles in the volume V. For this reason, $n^\mu(xyz)$ is the **particle density vector.**

10.4 Conservation of particle number

There is no need for the volume of integration to be flat. Imagine a further volume V' in spacetime, bounded by the *same* surface S (Fig. 10.2); we shall take V' to be **spacelike**, that is, every volume element $\mathrm{d}V'$ is a timelike covariant vector (but these vectors are no longer parallel). Every worldline embraced by S intersects V and V' *exactly once each*: N has the same value for V and for V'. Therefore the value of the integral is independent of the volume of integration; it depends only on the bounding surface S. (This is an example of a **conservation law**. Worldlines are assumed not to appear or to disappear in the spacetime between V and V'; particle number is conserved.)

Taken together, the two volumes V and V' form the boundary of an enclosed hypervolume H. By the 4D version of Gauss' Theorem, the difference of the integrals may be expressed as an integral over the intervening H; since the difference is to be zero, the result is

$$0=\int_H \partial_\mu n^\mu \,\mathrm{d}^4H.$$

The only way for this to be true whatever V' is chosen is that

$$\partial_\mu n^\mu = 0 \quad \text{throughout;} \tag{10.2}$$

the **divergence** of **n** is to be zero everywhere. This is the *differential version* of the conservation law.

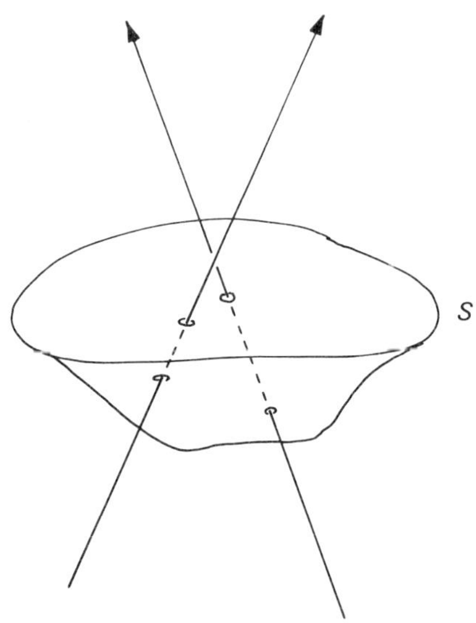

Fig. 10.2

In the specific Minkowski coordinates t, x, y, z, this may be written

$$\partial_t n^t = -\partial_x n^x - \partial_y n^y - \partial_z n^z \equiv -\nabla\cdot(n^x, n^y, n^z).$$

This has the characteristic shape of a **continuity equation**. The rate of change of the ordinary particle density n^t is related to the **particle flux** $\mathbf{n} = (n^x, n^y, n^z)$.

10.5 The matter tensor

Particle number is not the only interesting entity. A particle of restmass m and velocity $\mathbf{v}$ carries a contravariant **fourmomentum**

$$P^\mu = m\gamma(1, v^x, v^y, v^z) \quad \text{with} \quad \gamma = \frac{1}{\sqrt{(1 - v^2/c^2)}}. \tag{10.3}$$

We may ask for the *total* fourmomentum of all the particles whose worldlines are embraced by the surface S. (There is no problem in summing such vectors at different places in the *flat* spacetime of Special Relativity.) If the worldlines lie thickly enough we may introduce an appropriate density, in terms of which the result may be written as an integral over V:

$$P^\mu = \int_V T^{\mu\nu}\,\mathrm{d}V_\nu.$$

The new density $\mathbf{T}$ is the **energy–momentum–stress tensor** or, briefly, the **matter tensor** at $txyz$. It is the terse summary of the energy–momentum densities and currents in a material medium whose detailed motions may be very complicated. Its generalization to General Relativity is of the utmost importance, as it plays the part of the source of gravity.

In Special Relativity, each component of the total fourmomentum of an isolated system is a constant of the motion. (This is in fact conservation of energy and momentum, and is sometimes taken as the starting point of relativistic mechanics.) As before, we may examine the consequences for the two volumes V and V'; the conclusion is that an integral over an intervening hypervolume H is zero—

$$0 = \int_H \partial_\nu T^{\mu\nu}\,\mathrm{d}^4H,$$

leading to the differential version of the conservation of energy and momentum,

$$\partial_\nu T^{\mu\nu} = 0. \tag{10.4}$$

This conservation law will play a crucial part in later developments.

10.6 Interpreting the elements of the matter tensor

An observer stationary in the $txyz$-frame will interpret the various elements of $\mathbf{T}$ as follows. All his integrations over V will take place in his space of simultaneity $t = \text{constant}$, and the volume element is therefore

$$\mathrm{d}V_\mu = |\mathrm{d}V|(1, 0, 0, 0).$$

The total mass in V is then

$$P^t = \int T^{tt}\,\mathrm{d}V, \tag{10.5}$$

so that T^{tt} is interpreted as the **mass-density** (or energy-density if modified by a factor c^2). The mass referred to is not just the proper mass, but in general includes all kinds of energy, whatever its origin. The total momentum in V is

$$(P^x, P^y, P^z) = \int (T^{xt}, T^{yt}, T^{zt})\,\mathrm{d}V, \tag{10.6}$$

so that the 3-vector (T^{xt}, T^{yt}, T^{zt}) is the **momentum-density**.

The remaining twelve components may be interpreted using the conservation law, written out as

$$\partial_t T^{tt} = -\nabla\cdot(T^{tx}, T^{ty}, T^{tz})$$

and

$$\partial_t T^{it} = -\nabla\cdot(T^{ix}, T^{iy}, T^{iz}) \quad \text{for } i = x, y, z \text{ in turn.}$$

These have the form of four continuity equations from which, by inspection, we may infer the interpretations

$$\begin{aligned} \text{mass flux} &= (T^{tx}, T^{ty}, T^{tz}) \\ i\text{-momentum flux} &= (T^{ix}, T^{iy}, T^{iz}) \quad (i = x, y, z). \end{aligned}$$

The nine components of the momentum flux form the 3D **stress tensor**. As it happens, in conventional mechanics the stress tensor is invariably *symmetric*; this is necessary for conservation of angular momentum. Also, if this stress tensor remains symmetric when the full tensor **T** undergoes a Lorentz transformation, then the full tensor **T** must itself be symmetric:

$$T^{\mu\nu} = T^{\nu\mu}. \tag{10.7}$$

10.7 Some special instances of the matter tensor

(i) **Motionless dust** A *dust* is a collection of particles with straight, mutually parallel worldlines. If the worldlines are parallel to the t-axis of a certain inertial observer, then for him the dust is *motionless*. In this case, there is no momentum, and there are no flows of any kind. Only one element of **T** is nonzero:

$$T^{tt} = \rho_0 = \text{restmass density.}$$

Usually the dust will be taken to be *uniform* (ρ_0 is everywhere the same).

(ii) **Dust with fixed velocity v** for the motionless dust just considered, the contravariant four-velocity is $v^\mu = (1, 0, 0, 0)$, allowing us to write

$$T^{\mu\nu} = \rho_0 v^\mu v^\nu.$$

Since this has the form of a correct tensor equation, it preserves its shape under Lorentz transformation, and therefore provides the matter tensor for moving dust.

In fact, with $v^\mu = \gamma(1, v^x, v^y, v^z)$,

$$T^{\mu\nu} = \rho_0 \gamma^2 \begin{bmatrix} 1 & v^x & v^y & v^z \\ v^x & (v^x)^2 & v^x v^y & v^x v^z \\ v^y & v^y v^x & (v^y)^2 & v^y v^z \\ v^z & v^z v^x & v^z v^y & (v^z)^2 \end{bmatrix}.$$

Here there is a nonzero momentum, a nonzero mass-current, and a nonzero stress. The last will be perceived as a combination of a pressure and a shear on any absorbing barrier placed in the way of the particles.

(iii) **Monoenergetic gas** Here the particles all move with the same speed v, but their directions of motion are random; the particles are assumed not to interact in any way. Such a gas may be regarded as a collection of moving dusts—one for each direction—and the matter tensor is obtained by averaging the previous one over all directions. When this is done, elements like v^x, $v^x v^y$, etc. average to zero, while the average of $(v^x)^2$ etc. is $v^2/3$. We obtain

$$T^{\mu\nu} = \rho_0 \gamma^2 \begin{bmatrix} 1 & \cdot & \cdot & \cdot \\ \cdot & v^2/3 & \cdot & \cdot \\ \cdot & \cdot & v^2/3 & \cdot \\ \cdot & \cdot & \cdot & v^2/3 \end{bmatrix}.$$

Here we see the appearance of an isotropic **pressure**

$$p = \tfrac{1}{3}\rho_0 \gamma^2 v^2.$$

(iv) **Stationary general gas** Again the particles are to be noninteracting, but now they have a distribution of speeds v. Averaging the matter tensor for the monoenergetic gas over this distribution gives

$$T^{\mu\nu} = \begin{bmatrix} \rho & \cdot & \cdot & \cdot \\ \cdot & p & \cdot & \cdot \\ \cdot & \cdot & p & \cdot \\ \cdot & \cdot & \cdot & p \end{bmatrix}$$

where ρ and p are the averaged mass-density and pressure. The relation between the two depends on the shape of the speed distribution; when there is statistical equilibrium, this relation is normally given in the form of an equation of state for the gas.

With $V^\mu = (1, 0, 0, 0)$ and $g_{\mu\nu} = \mathrm{diag}(c^2, -1, -1, -1)$, we may write

$$T^{\mu\nu} = (\rho + p/c^2) V^\mu V^\nu - p g^{\mu\nu}.$$

Since this has a correct tensor form, it defines **T** for an observer moving past the gas with any other four-velocity V.

(v) An important case of the foregoing is **radiation**. It is a general result for radiation in equilibrium that

$$\text{radiation pressure} = (\text{energy-density})/3.$$

In this case, therefore, the matter tensor is

$$T^{\mu\nu} = \tfrac{4}{3}\rho V^{\mu} V^{\nu} - \tfrac{1}{3}\rho c^2 g^{\mu\nu}$$
$$= \begin{bmatrix} \rho & \cdot & \cdot & \cdot \\ \cdot & \tfrac{1}{3}\rho c^2 & \cdot & \cdot \\ \cdot & \cdot & \tfrac{1}{3}\rho c^2 & \cdot \\ \cdot & \cdot & \cdot & \tfrac{1}{3}\rho c^2 \end{bmatrix}.$$

10.8 Compact objects

We shall define a **compact object** as something whose constituent parts never stray far from an average worldline, and which is surrounded by empty space. A solid object in free fall, such as an artificial satellite, or the planet Mercury, is an example. The object need not be solid, provided it is gravitationally held together strongly enough to remain in one piece—the Sun, for example. The Solar System as a whole will do, or even our entire Galaxy. The requirements are that the object should be isolated—at least approximately—and that it should remain inside an appropriate worldtube throughout its history (Fig. 10.3).

The conservation law for the matter tensor has important consequences for compact objects in Special Relativity. Any integral over a surface S completely enclosing the object (and therefore entirely in empty space) vanishes if its integrand contains any part of the matter tensor as a factor. Such an integral may always be transformed to an integral over the enclosed volume V by the Gauss Theorem.

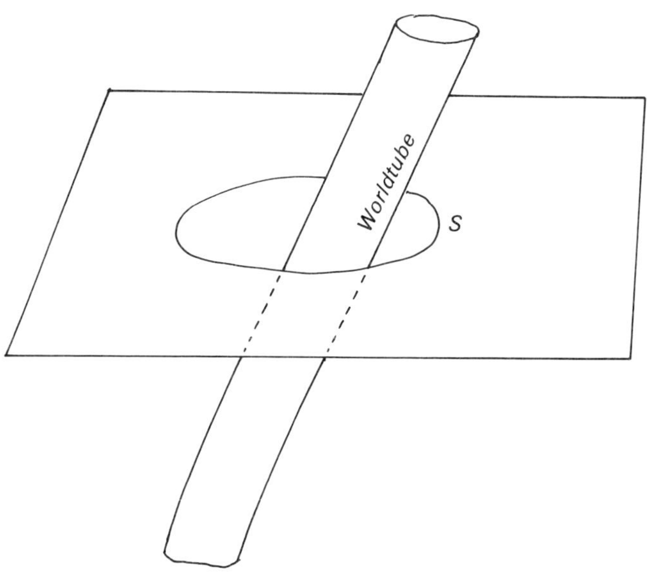

Fig. 10.3

We work entirely in the space $t = \text{constant}$ and write for convenience

$$\begin{aligned} \rho &= T^{tt} = \text{mass-density} \\ p^i &= T^{ti} = \text{momentum-density} = \text{mass flux} \\ t^{ij} &= T^{ij} = \text{stress} = \text{momentum flux} \end{aligned}$$

where i and j range over the 3D Cartesian values x, y, z.

There are four particularly important instances.

(a) When the continuity equation

$$\nabla \cdot \mathbf{p} \equiv \partial_i p^i = -\partial_t \rho \quad \text{(summed, of course, over } i = x, y, z\text{)}$$

is integrated through V, the divergence on the left yields zero, by the Gauss Theorem. Hence

$$0 = -\mathrm{d}M/\mathrm{d}t,$$

where

$$M = \int_V \rho \, \mathrm{d}V. \tag{10.8}$$

The total mass M of a compact object is constant. It may be worth pointing out that this is not quite as obvious as it may seem. In this context 'mass' is to include everything that is identifiable as mass or energy, even potential energy.

(b) The other continuity equation

$$\partial_i t^{ij} = -\partial_t p^j$$

may be integrated similarly, giving

$$0 = -\mathrm{d}P^j/\mathrm{d}t,$$

where

$$P^j = \int_V p^j \, \mathrm{d}V. \tag{10.9}$$

The total momentum (P^x, P^y, P^z) is constant.

(c) A more complicated divergence is

$$\begin{aligned} \partial_i(p^i r^j) &= (\partial_i p^i) r^j + p^i (\partial_i r^j) \\ &= -(\partial_t \rho) r^j + p^j \end{aligned}$$

where the first continuity equation has been used. (We have written r^x for x, etc.) Integrating gives

$$0 = M \frac{\mathrm{d}R^j}{\mathrm{d}t} - P^j,$$

where

$$R^j = \frac{1}{M} \int_V \rho r^j \, \mathrm{d}V. \tag{10.10}$$

The velocity of the centre of momentum R is constant. This of course gives a geodesic worldline in Minkowski space; we need this for consistency.

(d) The next instance uses the symmetry of the stress tensor. We have

$$\begin{aligned}\partial_i(t^{ij}r^k) &= (\partial_i t^{ij})r^k + t^{ij}(\partial_i r^k)\\ &= -(\partial_t p^j)r^k + t^{kj}.\end{aligned}$$

Continuity equations have again been used. Integrating over V gives

$$0 = -\frac{\mathrm{d}}{\mathrm{d}t}\int_V p^j r^k\,\mathrm{d}V + \int_V t^{kj}\,\mathrm{d}V.$$

Since $\mathbf{t}$ is symmetric, $t^{kj} = t^{jk}$, it follows that

$$\frac{\mathrm{d}J^{jk}}{\mathrm{d}t} = 0,$$

where

$$J^{jk} = \int (r^j p^k - p^j r^k)\,\mathrm{d}V. \tag{10.11}$$

The total angular momentum (J^{yz}, J^{zx}, J^{xy}) is constant.

10.9 The move to General Relativity

We have unveiled the matter tensor $\mathbf{T}$, and have shown how well fitted it is to describe the distribution of matter in Minkowski spacetime. It generates all the desirable features, such as conservation of energy, momentum, and angular momentum, and ensures that the centre of mass of a compact object follows a geodesic worldline. In the spirit of the Principle of Equivalence ('anything in the Special has its counterpart in the General'), we claim that in General Relativity the distribution of matter will continue to be described by the tensor $\mathbf{T}$. There is one slight, but important, change: the conservation law must now use the *covariant* derivative, not the ordinary derivative,

$$T^{\mu\nu}{}_{;\nu} = 0. \tag{10.12}$$

Remarkably, this is now *not* a conservation law. Written out explicitly, it becomes (see section 6.1)

$$\partial_\nu T^{\mu\nu} + \Gamma^\mu{}_{\alpha\nu}T^{\alpha\nu} + \Gamma^\nu{}_{\alpha\nu}T^{\alpha\mu} = 0. \tag{10.13}$$

The Gauss Theorem applies only to part of the left side. The terms which involve the Γs are related to the *kinematic forces* acting on the matter distribution in the usual way; they are the all-pervasive gravity forces. Their presence spoils the shape of the continuity equations. In General Relativity there is no universal conservation.

This is not the loss it may seem. In mechanics, all conservation has its roots in an underlying symmetry or invariance. For example, the Schwarzschild solution is invariant against rotation and displacement in time ('spherically symmetric', 'static') but not displacement in space. Consequently, the motions of planets possess certain first integrals, associated with conservation of energy and angular momentum, but there are no integrals for linear momentum. A partial symmetry leads only to a partial conservation. A general spacetime with no symmetry displays no conservation.

On the other hand, the usual conservation laws still apply in a local sense for an inertial observer in his smallish laboratory. In his local geodesic coordinates (sections 5.10 and 5.11), all the Γs are very nearly zero, and thus to a high degree of approximation the 'ordinary' divergence is zero. Consequently, within certain limits, he can safely use all the conservation laws he has become accustomed to.

10.10 The Bianchi identities

For a short time, Einstein believed that the Ricci tensor and the matter tensor were to be equated; in this way, matter would become the 'source' of gravity. This does not work, since the divergence of the matter tensor is desired to be zero, and that of the Ricci tensor almost always is not.

As it happens, there exists a tensor, closely related to the Ricci tensor, whose divergence is *identically* zero. To find it, we start with the **Bianchi identities**, which are universally satisfied by the covariant derivative of the Riemann tensor,

$$R_{\mu\nu\rho\sigma;\tau} + R_{\mu\nu\sigma\tau;\rho} + R_{\mu\nu\tau\rho;\sigma} = 0.$$

A reason for supposing that such identities must exist has been given in section 7.5 (and see also Problem 3 of that chapter). They form a set of 1024 equations, most of which say nothing; there are essentially only 24 identities of a nontrivial kind.

This set of identities may be dramatically collapsed by multiplying by $g^{\mu\rho}g^{\nu\sigma}$ and summing. Using the definitions of the Ricci tensor and the scalar curvature, along with the antisymmetries of the Riemann tensor (sections 6.4 and 6.5, 8.2 and 8.3), we are left with the four equations

$$R_{\mathrm{S};\tau} - R^{\rho}{}_{\tau;\rho} - R^{\sigma}{}_{\tau;\sigma} = 0;$$

this may be written

$$(R_{\mathrm{S}} g^{\sigma}{}_{\tau} - 2R^{\sigma}{}_{\tau})_{;\sigma} = 0.$$

Define the **Einstein tensor G** (for **G**ravity)

$$G^{\mu\nu} = R^{\mu\nu} - \tfrac{1}{2} g^{\mu\nu} R_{\mathrm{S}}. \tag{10.14}$$

Then the tensor **G** satisfies, *as a matter of identity*, the divergence condition

$$G^{\mu\nu}{}_{;\nu} = 0. \tag{10.15}$$

10.11 The gravity field equations in the presence of matter

We now have two tensors: **T**, whose divergence we should like to be zero, and **G**, whose divergence is inevitably zero. The temptation to identify them is irresistible, and so we obtain the final version of the field equations for gravity,

$$G^{\mu\nu} = 8\pi(G/c^2)T^{\mu\nu}. \tag{10.16}$$

The choice of the constant multiplier will be justified later. The G on the left is the Einstein tensor, while the G on the right is the gravitational constant; this never seems to cause confusion in practice!

The equation (10.16) has much to commend it. If it is adopted, then—at least in an approximate local sense—energy, momentum, and angular momentum are all obliged to be conserved, and the fact that free-fall worldlines are geodesics need no longer be brought in as an extra requirement. Within the confines of the Einstein theory there has been just one competitor, the one with an extra cosmological term; this will be considered in Chapter 11 in the context of cosmology.

10.12 Geometrized units

Frequently, **geometrized units** are chosen such that $c = 1$ and the gravity constant $G = 1$. Mass then has dimensions of length, and mass-density dimensions of $(\text{area})^{-1}$. Since curvature also has dimensions $(\text{area})^{-1}$, the field equation with the source term is dimensionally satisfactory.

In geometrized units, all physical quantities acquire dimensions which take the form $(\text{length})^n$; some of the more common are given in Table 10.1. The change from geometrized to SI units is achieved by *multiplying* by the appropriate conversion factor, with

$$c = 2.998 \times 10^8 \,\text{m}\,\text{s}^{-1},$$
$$G = 6.67 \times 10^{-11} \,\text{N}\text{m}^2\,\text{kg}^{-2}.$$

In order that the geometrized units should be unambiguous for the components of vectors and tensors, an associated scalar needs to be found. For vectors, $\surd(A^\mu A_\mu)$ will do. For tensors with two affixes, there is a choice: we may use the scalar $\surd(T^{\mu\nu}T_{\mu\nu})$, or we may use the scalar eigenvalues of the mixed tensor $T^\mu{}_\nu$; each leads to the same determination of units. Indeed, the mixed tensor usually gives the version most closely related to physical magnitudes, as we have already seen for the Riemann tensor and the tidal tensor $(R^{\mu\nu}{}_{\rho\sigma}, \Delta^\mu{}_\nu)$.

Throughout the rest of this book, geometrized units will be used.

Table 10.1 Relation between SI and geometrized units

	Dimensions: SI units	Dimensions: Geometrized units	Conversion factor
Mass (kg)	M	L	c^2/G
Length (m)	L	L	1
Time (s)	T	L	c^{-1}
Velocity ($\text{m}\,\text{s}^{-1}$)	LT^{-1}	1	c
Acceleration ($\text{m}\,\text{s}^{-2}$)	LT^{-2}	L^{-1}	c^2
Tide	T^{-2}	L^{-2}	c^2
Momentum ($\text{kg}\,\text{m}\,\text{s}^{-1}$)	MLT^{-1}	L	c^3/G
Energy (J)	ML^2T^{-2}	L	c^4/G
Power ($\text{J}\,\text{s}^{-1}$)	ML^2T^{-3}	1	c^5/G
Force (N)	MLT^{-2}	1	c^4/G
Mass-density ($\text{kg}\,\text{m}^{-3}$)	ML^{-3}	L^{-2}	c^2/G
Momentum-density	$ML^{-2}T^{-1}$	L^{-2}	c^3/G
Energy-density ($\text{J}\,\text{m}^{-3}$)	$ML^{-1}T^{-2}$	L^{-2}	c^4/G
Stress, pressure (P)	$ML^{-1}T^{-2}$	L^{-2}	c^4/G

10—Notes and problems

1. Show that, in 4D only, in any system of coordinates *uvwt*,

$$G^u{}_u = -R^{vw}{}_{vw} - R^{vt}{}_{vt} - R^{wt}{}_{wt}, \text{ etc.,}$$

and

$$G^u{}_v = R^{uw}{}_{vw} + R^{ut}{}_{vt}, \text{ etc.}$$

Thus the mixed tensor **G** is easily written down when the mixed version of the Riemann tensor is known.

2. Show that, in 2D, the Einstein tensor *vanishes identically*. (Thus there is no hope for an Einstein theory in 2D.)

3. Verify

$$T^\mu{}_{\nu;\mu} \equiv \partial_\mu T^\mu{}_\nu + \Gamma^\mu{}_{\mu\alpha} T^\alpha{}_\nu - \Gamma^\alpha{}_{\nu\mu} T^\mu{}_\alpha.$$

(This mixed version of the conservation law will be used in section 12.2.)

4. The Gauss Theorem requires a little care in general coordinates $u, v, w \ldots$, since the volume element is not $\mathrm{d}u\,\mathrm{d}v\,\mathrm{d}w, \ldots$, but rather $\sqrt{g}\,\mathrm{d}u\,\mathrm{d}v\,\mathrm{d}w \ldots$, where g is the *determinant* of the metric matrix. It is necessary to remember that $\sqrt{g}$ is to be regarded as part of the integrand when the Gauss Theorem is applied. (For clarification, consider ordinary 3D polar coordinates $r\theta\phi$, for which

$$\mathrm{d}s^2 = \mathrm{d}r^2 + r^2\,\mathrm{d}\theta^2 + r^2 \sin^2\theta\,\mathrm{d}\phi^2.$$

Verify that $g = r^4 \sin^2\theta$, and that therefore the volume element is the familiar $r^2 \sin\theta\,\mathrm{d}r\,\mathrm{d}\theta\,\mathrm{d}\phi$.) Note that in a $(1+3)$D spacetime we prefer to use $\sqrt{(-g)}$ to avoid the entry of complex quantities.

Show that

$$\partial_\rho \sqrt{(-g)} = \tfrac{1}{2}\sqrt{(-g)}\,g^{\mu\nu}\partial_\rho g_{\mu\nu} = \sqrt{(-g)}\,\Gamma^\mu{}_{\mu\rho}.$$

(The first equality follows from the definition of a determinant, the second from the formula (5.2) for the Christoffel symbols.) Hence show that (10.13) may be rewritten

$$\partial_\nu\{T^{\mu\nu}\sqrt{(-g)}\} + \Gamma^\mu{}_{\alpha\nu} T^{\alpha\nu}\sqrt{(-g)} = 0.$$

This is now in the 'correct' form for applying the Gauss Theorem: if the second term is absent, we have a conservation law, otherwise there are apparently kinematic forces acting on the matter distribution. Refer to section 5.13 for a similar situation.

11

Cosmology

11.1 The smooth Universe

Our Universe is very irregular. Most of it is near-vacuum, but in some regions the mass-density rockets to enormous values. Any detailed treatment will therefore be very complicated.

Cosmology—at least in the first instance—is the study of an entire smoothed-out Universe. Local irregularities are replaced with an average matter tensor which is taken to be the same at all places. But the matter tensor is not expected to be the same at all *times*; it is necessary to allow for a nonstatic development to achieve anything of interest.

It is usual to consider only universes which satisfy this **cosmological principle** that—imprecisely—the entire Universe, viewed at a fixed time, is homogeneous and isotropic. This has a limited basis in experimental observation: it is true that from where we stand the Universe is isotropic to a high degree (apart from fairly local irregularities), and that what we can see of it seems on the whole to be reasonably homogeneous. Of course, we have no way of telling that the Universe is at all homogeneous and isotropic when seen from other places or at other times. To adopt the cosmological principle is therefore a matter of supposition. It has the great merit of keeping the mathematics simple.

11.2 The Robertson–Walker metric

The metric for a complete universe which embodies the cosmological principle takes the form

$$d\tau^2 = dt^2 - [R(t)]^2\, dS^2. \qquad (11.1)$$

The time-coordinate t is the **epoch**. At any constant epoch, the cosmological principle dictates that the Universe is to be homogeneous and isotropic; this is achieved by making dS^2 the metric for a homogeneous and isotropic 3D space. The only possibilities are spaces of constant curvature k which may be positive, zero, or negative. It is usual to restrict the possibilities to $k = +1, 0, -1$, since the magnitude of k may be absorbed in a redefinition of $R(t)$. (When $k = 1$, $cR(t)$ plays the part of the radius of the space part of the metric.) In general, R is called the **scale factor**. It is regrettable that the letter R has been universally used for this entity: it has nothing to do with the Riemann or Ricci tensors.

The 3D space of dS^2 will need three coordinates (u^1, u^2, u^3) to describe it. Note that there are no cross terms like $dt\, du^i$ in the Robertson–Walker metric; the presence of such a term would imply that the $+u^i$ direction and the $-u^i$ direction would look different: the metric would not have the required isotropy. We shall find it convenient to use, in the three different cases,

$$\mathbf{k} = +1 \quad dS^2 = d\chi^2 + \sin^2\chi(d\theta^2 + \sin^2\theta\, d\phi^2) \tag{11.2}$$

$$\mathbf{k} = 0 \quad dS^2 = d\chi^2 + \chi^2(d\theta^2 + \sin^2\theta\, d\phi^2) \tag{11.3}$$

$$\mathbf{k} = -1 \quad dS^2 = d\chi^2 + \sinh^2\chi(d\theta^2 + \sin^2\theta\, d\phi^2) \tag{11.4}$$

(see Chapter 6, Problems 6, 7). If $k = +1$, then the Universe is *closed*, with a finite volume; otherwise it is *open*, with infinite volume. In each case, the coordinate χ is a *radial* coordinate, while the coordinates θ and ϕ show a clear spherical symmetry. On account of the homogeneity, it is often possible, by a proper choice of the centre of coordinates, to reduce problems to the case $\theta = \text{const}$, $\phi = \text{const}$, and to work in terms of the coordinate χ alone.

The function $R(t)$ represents the *history* of the Universe. We have to ask two questions: How does the history affect what goes on? and: What determines the history itself?

11.3 Fundamental observers

It is easy to verify that any worldline of constant χ, θ, and ϕ is a geodesic, and that proper time along this worldline is essentially identical with the epoch t. An inertial observer with such a worldline is a **fundamental observer**. There is a sense in which he is 'at rest': he is the one who sees his environment as isotropic. Other inertial observers do not.

It is useful to ask how one fundamental observer views another. Place the centre of coordinates at one of them, and suppose that the radial coordinate of the position of the other is χ, which is of course constant. Then their separation (taken at constant epoch) is $cR(t)\chi$; this is not constant. Fundamental observers will perceive their neighbours as moving radially away from (or perhaps towards) them. Since the separation is strictly proportional to the coordinate χ, it is simplest to interpret the motion as due to an overall expansion or contraction of the space itself.

The early theoreticians viewed this result as thoroughly undesirable, and went to some lengths to circumvent it. However, at about that time, observations on the reddening of light from very distant objects suggested strongly that the actual Universe is undergoing a substantial expansion.

11.4 The cosmological redshift

Imagine that a fundamental observer A at the centre of coordinates $\chi = 0$ transmits, at epoch t_A, a pulse of light which is received by a second fundamental observer B at position χ_B and epoch t_B. On account of the spherical symmetry, along the null-worldline of the pulse, both θ and ϕ are constant; consequently—whatever the choice of k—

$$0 = dt^2 - R^2 d\chi^2,$$

which gives on integration

$$\int_{t_A}^{t_B} dt/R(t) = \chi_B.$$

If A emits a further pulse at a slightly later epoch $t_A + \Delta t_A$, and B receives it at $t_B + \Delta t_B$, then (since χ_B does not change)

$$\Delta \int_{t_A}^{t_B} \frac{dt}{R(t)} \equiv \frac{\Delta t_B}{R(t_B)} - \frac{\Delta t_A}{R(t_A)} = 0.$$

If we think of successive 'pulses' as the wave-crests of emitted radiation, we may take Δt_A to be the period of a photon transmitted by A, and Δt_B to be the period of the same photon as perceived by B. If the Universe is *expanding*, that is, if

$$R(t_B) > R(t_A),$$

then

$$\Delta t_B > \Delta t_A;$$

the period of the photon increases. This is the **cosmological redshift**.

If we assume—as most do—that atomic physics was the same in the past as it is today, then we shall believe in particular that the characteristic spectrum of, say, atomic hydrogen has not changed. Let us observe the emission spectrum of atomic hydrogen in a remote—and therefore earlier—part of the Universe, and compare it with the same spectrum produced in our local laboratory. The effect of the cosmological redshift will be *to displace one spectrum bodily with respect to the other*: each and every photon is reddened in the same proportion.

The observed reddening of light from distant galaxies was put on a firm observational basis in the late 1920s, notably by Edwin Hubble and Milton Humason. The theoretical suggestion that such a phenomenon might exist was hinted at by de Sitter as early as 1917, as a result of arguments from General Relativity.

The cosmological redshift has another effect of a rather different kind. If we ignore radiation from all the localized hot spots in the Universe, there remains a very nearly isotropic background of black-body radiation with no identifiable source. Any cosmological redshift preserves the Planck profile of such radiation; all it does is to change the temperature T according to $TR(t) = \text{constant}$. Currently, this cosmic thermometer registers 2.7 K, but the temperature must have been higher in the past if $R(t)$ is on the increase.

The black-body radiation is an insurmountable barrier to the observation of the very early Universe when R was much smaller than it is now. When the temperature of the background is high enough for easy interaction with atomic transitions—a few thousand K, say—objects will become invisible, having merged into the equilibrium; nothing will penetrate the mists. This is the epoch at which it can reasonably be said that the background radiation originated; it has been cooling ever since. (It has been fashionable to say that the 2.7 K background is the leftover of the *Big Bang*; for this reason it is sometimes referred to as **relic radiation**. However, it is rather more recent than that.)

11.5 A particle in free fall

The Lagrangian for a non-null geodesic worldline is

$$L = \dot{t}^2 - R^2\dot{\chi}^2 = 1.$$

(Without any real loss of generality, we have again been able to set $\dot{\theta}$ and $\dot{\phi}$ to zero, and thus consider radial motions only. On account of the 3D homogeneity and isotropy, other motions are essentially identical, but their representation is more complicated. And again, the work is independent of k.) There is a first integral of the motion,

$$R^2\dot{\chi} = \alpha \text{ (constant)}. \tag{11.5}$$

An elimination and rearrangement gives

$$\dot{t}^2 = 1 + \frac{\alpha^2}{R^2}, \tag{11.6}$$

whence, integrating,

$$\tau = \int_0^t \frac{R\,\mathrm{d}t}{\sqrt{(R^2+\alpha^2)}}.$$

Thus, when the history $R(t)$ is known, we have the relation between t and τ. (When $\alpha = 0$, $t = \tau$; this gives the worldline of a fundamental observer.) The solution is completed—after a further integration—by the relation between χ and τ.

At any instant, the velocity of the particle with respect to the background of fundamental observers is $v = R\,\mathrm{d}\chi/\mathrm{d}t$; this is

$$v = R\dot{\chi}/\dot{t} = \frac{\alpha}{\sqrt{(R^2+\alpha^2)}}$$

from (11.5) and (11.6). Thus if $R(t)$ is increasing (the Universe is expanding), v decreases: free fall motions become steadily slower as epoch passes.

11.6 The Christoffel symbols for the Robertson–Walker metric

Let us find the Christoffel symbols and the Riemann tensor for the case $k = 0$, and leave the other cases to the reader. The metric in question is given at (11.3) and, because the space part of the metric is flat, we may effect an enormous simplification by changing to *Cartesian* coordinates x, y, and z:

$$\mathrm{d}\tau^2 = \mathrm{d}t^2 - R^2(\mathrm{d}x^2 + \mathrm{d}y^2 + \mathrm{d}z^2),$$

and

$$L = \dot{t}^2 - R^2(\dot{x}^2 + \dot{y}^2 + \dot{z}^2).$$

Now (writing R' for $\mathrm{d}R/\mathrm{d}t$)

$$\frac{\mathrm{d}}{\mathrm{d}\tau}\left(\frac{\partial L}{\partial \dot{t}}\right) - \frac{\partial L}{\partial t} = \frac{\mathrm{d}}{\mathrm{d}\tau}(2\dot{t}) + 2RR'(\dot{x}^2 + \dot{y}^2 + \dot{z}^2)$$

whence, in the style of section 5.3,

$$F^t = \ddot{t} + RR'\dot{x}^2 + RR'\dot{y}^2 + RR'\dot{z}^2.$$

Similarly,

$$\frac{\mathrm{d}}{\mathrm{d}\tau}\left(\frac{\partial L}{\partial \dot{x}}\right) - \frac{\partial L}{\partial x} = \frac{\mathrm{d}}{\mathrm{d}\tau}(-2R^2\dot{x}) = -2R^2\ddot{x} - 4RR'\dot{t}\dot{x},$$

and

$$F^x = \ddot{x} + 2(R'/R)\dot{t}\dot{x}.$$

By the obvious symmetry between x and y and z, we may write without difficulty

$$F^y = \ddot{y} + 2(R'/R)\dot{t}\dot{y}, \quad F^z = \ddot{z} + 2(R'/R)\dot{t}\dot{z}.$$

We may now read off the nine nonzero Γs,

$$\begin{gathered} \Gamma^t{}_{xx} = \Gamma^t{}_{yy} = \Gamma^t{}_{zz} = RR', \\ \Gamma^x{}_{tx} = \Gamma^x{}_{xt} = \Gamma^y{}_{ty} = \Gamma^y{}_{yt} = \Gamma^z{}_{tz} = \Gamma^z{}_{zt} = R'/R. \end{gathered} \tag{11.7}$$

11.7 The Riemann tensor for the Robertson–Walker metric

The Γs are assembled into the customary matrices

$$\mathbf{\Gamma}_t = \begin{bmatrix} \cdot & \cdot & \cdot & \cdot \\ \cdot & R'/R & \cdot & \cdot \\ \cdot & \cdot & R'/R & \cdot \\ \cdot & \cdot & \cdot & R'/R \end{bmatrix}, \quad \mathbf{\Gamma}_x = \begin{bmatrix} \cdot & RR' & \cdot & \cdot \\ R'/R & \cdot & \cdot & \cdot \\ \cdot & \cdot & \cdot & \cdot \\ \cdot & \cdot & \cdot & \cdot \end{bmatrix},$$

$$\mathbf{\Gamma}_y = \begin{bmatrix} \cdot & \cdot & RR' & \cdot \\ \cdot & \cdot & \cdot & \cdot \\ R'/R & \cdot & \cdot & \cdot \\ \cdot & \cdot & \cdot & \cdot \end{bmatrix}, \quad \mathbf{\Gamma}_z = \begin{bmatrix} \cdot & \cdot & \cdot & RR' \\ \cdot & \cdot & \cdot & \cdot \\ \cdot & \cdot & \cdot & \cdot \\ R'/R & \cdot & \cdot & \cdot \end{bmatrix}.$$

It is sufficient to find $\mathbf{B}_{tx}$ and $\mathbf{B}_{xy}$, and to rely on the symmetry to complete the picture. A straightforward calculation gives

$$\mathbf{B}_{tx} = \partial_t\mathbf{\Gamma}_x - \partial_x\mathbf{\Gamma}_t + \mathbf{\Gamma}_t\mathbf{\Gamma}_x - \mathbf{\Gamma}_x\mathbf{\Gamma}_t = \begin{bmatrix} \cdot & RR'' & \cdot & \cdot \\ R''/R & \cdot & \cdot & \cdot \\ \cdot & \cdot & \cdot & \cdot \\ \cdot & \cdot & \cdot & \cdot \end{bmatrix}$$

and

$$\mathbf{B}_{xy} = \partial_x\mathbf{\Gamma}_y - \partial_y\mathbf{\Gamma}_x + \mathbf{\Gamma}_x\mathbf{\Gamma}_y - \mathbf{\Gamma}_y\mathbf{\Gamma}_x = \begin{bmatrix} \cdot & \cdot & \cdot & \cdot \\ \cdot & \cdot & R'^2 & \cdot \\ \cdot & -R'^2 & \cdot & \cdot \\ \cdot & \cdot & \cdot & \cdot \end{bmatrix}.$$

From these we may read off

$$R^t{}_{xtx} = RR'', \quad R^x{}_{ttx} = R''/R, \quad R^x{}_{yxy} = R'^2, \quad R^y{}_{xxy} = -R'^2.$$

Raising one suffix and completing the picture by the 3D symmetry gives finally

$$R^{tx}{}_{tx} = R^{ty}{}_{ty} = R^{tz}{}_{tz} = -R''/R, \quad R^{xy}{}_{xy} = R^{xz}{}_{xz} = R^{yz}{}_{yz} = -R'^2/R^2.$$

The working for the metrics (11.2) and (11.4) is very similar, though more involved. The result is, for *any* k,

$$R^{t\chi}{}_{t\chi} = R^{t\theta}{}_{t\theta} = R^{t\phi}{}_{t\phi} = -R''/R, \quad R^{\chi\theta}{}_{\chi\theta} = R^{\chi\phi}{}_{\chi\phi} = R^{\theta\phi}{}_{\theta\phi} = -(R'^2 + k)/R^2. \quad (11.8)$$

There are a further 18 nonzero elements which are obtained via the antisymmetries of the tensor. On account of the 3D homogeneity and isotropy of the metrics (11.2)–(11.4) we may replace the coordinates χ, θ, ϕ by any coordinates we may care to choose; the expressions (11.8) for the mixed version of the Riemann tensor remain unchanged (though this cannot be said of the other versions). They are in fact the principal curvatures of the metric.

The Einstein tensor **G** follows immediately (see Chapter 10, Problem 1); its nonzero components are

$$G^t{}_t = 3(R'^2 + k)/R^2 \qquad = 8\pi\rho \quad (11.9)$$

$$G^\chi{}_\chi = G^\theta{}_\theta = G^\phi{}_\phi = 2R''/R + (R'^2 + k)/R^2 = -8\pi p. \quad (11.10)$$

(see (10.16); geometrized units have been used.) The Robertson–Walker universe contains a uniform distribution ρ of mass, under isotropic pressure p.

11.8 The Hubble constant and the deceleration parameter

In relating theory to the results of observation it is useful to have certain parameters. The first arises as follows. The distance X between two fundamental observers A and B, measured at constant epoch t, is $X = R(t)\chi$. The rate of change due to the expansion is

$$\begin{aligned} \mathrm{d}X/\mathrm{d}t &= R'\chi = R'X/R \\ &= HX \end{aligned}$$

where the so-called **Hubble constant** H is R'/R: it is the proportionate rate of change of separation. (In fact, H is not necessarily constant as epoch passes, since it depends on $R(t)$.)

The Hubble constant gives a first estimate of the reddening of light from a distant source; it gives an approximate value $\Delta\lambda/\lambda$ for unit distance (λ is wavelength). The reason why this cannot be more than approximate is that the reddening involves values of R at more than one epoch, whereas H does not. It is nevertheless the useful parameter for connecting theory with observation. Over sixty years astronomers have expended much effort in trying to obtain a good value for H; measuring a redshift is easy, and the main difficulty lies in being sure about the determination of the distance from a remote part of the Universe. The uncertainty involves a factor of about two; the accepted value is about $H = 8 \times 10^{-27}\,\mathrm{m}^{-1}$.

Rearrange (11.9) to obtain

$$\rho = \rho_c + 3k/8\pi R^2,$$

where $\rho_c = 3H^2/8\pi$. If $\rho > \rho_c$ (the *critical density*), then k is positive and the Universe is closed; otherwise the Universe is of infinite extent. It is therefore of great interest to determine the mass-density of our actual Universe. Unfortunately, the value is very uncertain, though it seems likely to be less than critical, suggesting that the Universe may actually be open. The accepted value of H leads to $\rho_c = 8 \times 10^{-54}\,\mathrm{m}^{-2}$ ($= 10^{-26}\,\mathrm{km\,m}^{-3}$ in SI units), while the observed density is hardly more than a few percent of this. However, it is very difficult to be sure that nothing is being missed.

The second parameter is the dimensionless **deceleration parameter**

$$q = -R''R/R'^2 = -(R''/R)H^{-2}. \tag{11.11}$$

Eliminating R''/R and $H = R'/R$ from (11.9)–(11.11) gives

$$q = \frac{\rho + 3p}{2\rho - 3k/4\pi R^2}.$$

There is no doubt that the present Universe is *matter-dominated*: its pressure is very small compared to the density. If we suppose that $p = 0$, then it is evident that the Universe is open or closed according as $q < 1/2$ or $q > 1/2$. This provides an alternative observational test which is independent of any measurements of density. So far, there are no estimates of q reliable enough to settle the matter.

11.9 The nonconservation of energy

If an 'equation of state' connecting ρ and p is available, we may write down a differential equation for $R(t)$ whose solution displays the entire history of a conceivable universe. In general, the solution will need to be obtained numerically, perhaps by computer. There are some interesting cases which can be solved exactly with ease; two of them are particularly important in that they are in a sense at opposite extremes, and a guess at intermediate behaviour is straightforward.

We shall first show that it is the *adiabatic* equation of state which needs to be used to relate ρ and p. Envisage a gas enclosed in a volume V at uniform pressure p and uniform energy density ρ; the total energy of the system is ρV. On an adiabatic change of volume, the change in energy is the work done at the boundary by the pressure, $\mathrm{d}(\rho V) = -p\,\mathrm{d}V$, or

$$V\,\mathrm{d}\rho + (\rho + p)\,\mathrm{d}V = 0.$$

An adiabatic change which proceeds smoothly as time passes will thus be governed by

$$\dot{\rho} + (\rho + p)(\dot{V}/V) = 0; \tag{11.12}$$

this represents *conservation of energy*, in the usual sense of thermodynamics, in the absence of heat transfer.

A very similar relation is obliged to hold in any of our cosmological models, as a result of the Einstein equations. Differentiating (11.9) gives

$$6R''R'/R^2 - 6(R'^2 + k)R'/R^3 = 8\pi\rho',$$

and a use of (11.9), (11.10) shows that

$$\rho' + (\rho + p)(3R'/R) = 0. \tag{11.13}$$

Now any volume V which expands in step with the cosmology goes as R^3, and therefore $V'/V = 3R'/R$, and we are led directly back to the adiabatic requirement.

Oddly, this does not represent cosmological conservation of energy, since there are no cosmological boundaries for the pressure to do work on. An expanding universe loses energy, but it loses it to nowhere in particular. However, there is really no reason to expect energy conservation, on two counts: conservation of energy is always associated with systems in a static environment, and such an environment is not provided by an expanding cosmology; and anyway, integration of one element of the matter tensor to yield a total energy is an ill-defined procedure in a curved space.

11.10 The dust-filled universe

Dust is material throughout which the pressure is zero. It is convenient (though not necessary) to think of it as a distribution of small granules separated by empty space. 'Small' means small enough for the distribution to be essentially continuous on the scale of the environment. From the standpoint of cosmology, the galaxies themselves are quite small enough in this sense, and a universe with a uniform distribution of dust is likely to provide a reasonably good picture of the Universe as we see it now.

Since the pressure p is everywhere zero, the adiabatic requirement collapses to

$$\rho'/\rho + 3R'/R = 0.$$

This integrates immediately to

$$\rho = (3/8\pi)A/R^3, \tag{11.14}$$

where A is a constant. Since the energy includes nothing apart from the restmass of the dust particles, we should expect that the energy-density should vary as the inverse cube of the scale factor. A positive energy calls for $A > 0$.

Substituting for (11.14) in (11.9) and rearranging gives

$$R'^2 = (A - kR)/R, \tag{11.15}$$

from which $R(t)$ may be found exactly (Problem 2). A qualitative idea of the shape of the solution may be gained from an approach already familiar from our work with orbits: we note that the right side of the last equation must be nonnegative. There are three cases.

$\boldsymbol{k = +1}$ The solution starts from a singularity—the **Big Bang**—at $R = 0$ (the energy-density is infinite) and rises to a maximum at $R = A$. Thereafter it falls back to the **Big Crunch** at $R = 0$. (Thus a spatially closed universe has a finite history.)

$\boldsymbol{k = 0}$ The solution starts from a Big Bang at $R = 0$ as before, but now rises to infinity, where the expansion rate falls to zero. This is the 'critical case'.

$\boldsymbol{k = -1}$ As for $k = 0$, except that the ultimate expansion rate is not zero.

(There are alternatives in the last two cases, where R falls from infinity to zero. These yield ever-contracting universes, the ultimate fate of which is a Crunch. However, all the evidence indicates that our Universe is expanding.)

11.11 The radiation-filled universe

Let us now go to the opposite extreme of the *radiation-dominated* universe. It is believed that our actual Universe was much smaller and hotter in the past, to the extent that the pressure had an important influence; in the earliest epochs, the energy and pressure of radiation would have been far more important than those of matter.

The equation of state for pure radiation is

$$\text{pressure} = (\text{energy-density})/3,$$

leading to the adiabatic requirement

$$\rho' + (4\rho/3)3R'/R = 0.$$

Integrating leads to

$$\rho = (3/8\pi)A^2R^{-4}, \tag{11.16}$$

where the positive constant A is disposable. The energy-density decreases more rapidly than the R^{-3} expected from the mere dilution as the space expands; the extra R^{-1} takes account of the cosmological redshift already calculated in section 11.4. This redshift afflicts every photon—and therefore the radiation as a whole—in the same way.

Inserting (11.16) in (11.9) leads to

$$R'^2 = (A^2 - kR^2)/R^2, \tag{11.17}$$

which can be solved exactly. For positive values of R (negative values are not relevant), the qualitative features of the possible solutions are so similar to the solutions to (11.15) that the comments of the last section may be taken word for word for the present case. This suggests what is in fact true: *any model expanding universe with positive energy density and positive pressure starts at a singularity where* $R = 0$. This is the celebrated cosmological **Big Bang**.

The energy-density of black-body radiation is related to temperature by Stefan's law,

$$\rho = \sigma T^4.$$

Comparison with (11.16) confirms the result of section 11.4, that $T = \text{constant}/R$. We need not be surprised that Stefan's law is so near the surface: the law is usually derived by considering the expansion of a radiation-filled enclosure. Here we are thinking of what is unquestionably the largest enclosure that can exist: the entire Universe.

11.12 The Einstein universe

In the early days of cosmology in the light of General Relativity, it was believed that any reasonable model would need to be static: it should not change with the passage of epoch. Einstein was well aware that his original equations were incapable of producing such a model, and introduced his **cosmological constant** Λ into the equations:

$$G^{\mu}{}_{\nu} - \Lambda g^{\mu}{}_{\nu} = 8\pi T^{\mu}{}_{\nu}.$$

As an example, consider a dust-filled universe with $k = 1$. The requirement that the pressure is zero is now

$$2R''/R + (R'^2 + 1)/R^2 - \Lambda = 0.$$

Provided Λ is positive, there is an obvious constant solution $R = R_0$, where $\Lambda = R_0{}^{-2}$, so that a static solution is now possible. Integrating as before now gives

$$RR'^2 + R - R^3/3R_0{}^2 = A \text{ (constant).}$$

The value of A for the static solution is $2R_0/3$, and the equation becomes

$$(RR')^2 = R(R - R_0)^2(R + 2R_0)/3R_0{}^2.$$

A glance at the graph of the right side shows that the static solution $R = R_0$ is unstable: one would expect rather to find an infinitely continued expansion or an ultimate collapse. As the universe was likely to be expanding anyway, the cosmological term was soon regarded as an extra complication which might be dispensed with. However, a few workers, notably Eddington, wanted to keep it, feeling that its presence would provide a link between cosmology and fundamental particles (the very large and the very small). The idea has not carried much conviction for the majority.

11.13 The Bondi–Gold–Hoyle universe

Even without the cosmological constant it is possible to have a universe which, though *not static*, nevertheless looks the same at all times. This remarkable universe is obtained by making the energy-density and the pressure independent of the epoch; also $k = 0$. This leads to

$$3R'^2/R^2 = 3H^2 \text{ (constant)} \quad \text{and} \quad R''/R = \text{constant.}$$

The solution is

$$R = \text{constant.} \exp Ht.$$

H is in fact the Hubble constant (which in this case really is a constant), while the deceleration parameter $q = -1$ at all times. The pressure is negative, so that the universe is under tension.

How is it that a universe can be expanding, and yet not change in appearance? Suppose, for example, that it is filled with a gas of massive particles: expansion must cause the particle restmass density to decrease. (The energy-density remains constant on account—naively—of the compensating potential energy increase in the tension as the universe stretches.) Surely the ever-increasing scarcity of particles will be noticed?

The bold line of argument, followed by Hermann Bondi, Thomas Gold and Fred Hoyle, was to allow for the **spontaneous creation** of new particles at the rate needed to keep the particle density constant; the mechanism would be driven by the potential energy of the tension made available by the expansion. Within the context of this book, there is no such mechanism; we have to call on the resources of quantum field theory to understand how such a thing can happen.

It is fair to say that steady state pictures like Einstein's and Hoyle's are not fashionable. The preference for Big Bang pictures may be as much a consequence of philosophical and psychological leanings as of the rather inconclusive evidence of observation. The Big Bang is more exciting, more spectacular, more useful for the popularizers than the more pedestrian unchanging steady state. Honestly, I do not think anyone is really sure of the truth of the matter.

11.14 The de Sitter universe

It is simple to calculate, from (11.8), the six principal curvatures for the Bondi–Gold–Hoyle (BGH) metric:

$$d\tau^2 = dt^2 - e^{2Ht}\{d\chi^2 + \chi^2(d\theta^2 + \sin^2\theta\, d\phi^2)\}; \tag{11.18}$$

they are all H^2, at all places and times. Apart from scale, there is only one such space with signature (1 + 3)D. The work of Problem 5 shows that an alternative metric for this spacetime is the de Sitter metric

$$d\tau^2 = (1 - H^2r^2)dT^2 - \frac{dr^2}{1 - H^2r^2} - r^2(d\theta^2 + \sin^2\theta\, d\phi^2). \tag{11.19}$$

This was obtained as one instance of a spherically symmetric *static* metric (Problem 3, Chapter 8). From the mathematical point of view, it is simple enough to show that these metrics are equivalent. From the physical point of view, how can they be reconciled? One seems to be static, the other not.

It is important to remember that coordinates can be thoroughly misleading. What we actually see as observers of the Universe (whatever its shape really is) is not determined by the coordinates at all, but by the worldlines of the signals—usually electromagnetic radiation—that arrive in our telescopes. Of course, we will be wise to choose 'helpful' coordinates for our calculations for the sake of convenience. But, convenient or not, the calculations must lead to the same *physical* predictions, whatever the coordinates used.

For example, the BGH universe is expanding, and distant fundamental observers will appear to be redshifted. Calculation shows that, in the static version of the metric, a distant fundamental observer is moving away from the origin ($dr/d\tau > 0$), and this explains part of the redshift as a Doppler shift. Moreover, the coefficient of dT^2 (namely $1 - H^2r^2$) is less than its value at the origin, giving a further gravitational redshift. The separation of the redshifts into two types is not very meaningful, arising as it does from the choice of coordinates. Observationally, there is a redshift, and that is all that ought to be said.

Perhaps the de Sitter universe provides us with rather too many possibilities. On account of its extreme regularity—shown by all six principal curvatures being equal and constant—it may be represented by a Robertson–Walker metric in many different ways: some give expanding universes; some give contracting ones (see Problem 6 for an example). However, since the equation of state is the most improbable $p = -\rho$, we feel entitled to set these models on one side.

Since a more sensible equation of state does not lead to the equality of all six principal curvatures, the corresponding Robertson–Walker metric is unique, and therefore unambiguous.

11—Notes and problems

1. For a fundamental observer, the four-velocity is (1, 0, 0, 0). Show that the corresponding tidal tensor is

$$\mathbf{\Delta} = \begin{bmatrix} \cdot & \cdot & \cdot & \cdot \\ \cdot & R''/R & \cdot & \cdot \\ \cdot & \cdot & R''/R & \cdot \\ \cdot & \cdot & \cdot & R''/R \end{bmatrix}$$

Hence show that if $\rho + 3p > 0$ a fundamental observer experiences an isotropic tidal compression.

2. In (11.15) use the substitution $R = A/(q^2 + k)$ to obtain

$$\frac{2A\,\mathrm{d}q}{(q^2+k)^2} = \mathrm{d}t.$$

Hence show that the three cases may be parametrically represented by

$$k=1: \quad R = \frac{A}{q^2+1}, \quad t = A\left(\frac{q}{q^2+1} + \arctan q\right), \quad -\infty < q < +\infty;$$

$$k=0: \quad R = \frac{A}{q^2}, \quad t = \frac{-2A}{3q^3}, \quad -\infty < q < 0;$$

$$k=-1: \quad R = \frac{A}{q^2-1}, \quad t = A\left(\frac{-q}{q^2-1} + \frac{1}{2}\ln\frac{q+1}{q-1}\right); \quad -\infty < q < -1.$$

Show that in the case $k = 1$, the maximum 'radius' of the Universe is A, and the total 'lifetime' (that is, the total passage of epoch) is πA. (Incidentally, you may like to show that the graph of R against t is a cycloid, traced out by a point fixed to the circumference of a circle of diameter A which rolls along the t-axis. This fact has negligible cosmological significance.)

3. The equation (11.16) is simpler. When k is not zero, use the replacement $v^2 = A^2 - kR^2$ to obtain $\mathrm{d}v = k\,\mathrm{d}t$ and $kR^2 = A^2 - (kt)^2$. When $k = 0$, this approach fails, but the problem is then even simpler. Show that the three cases are

$$k=1: \quad R = \sqrt{(A^2 - t^2)}, \quad -A < t < +A;$$

$$k=0: \quad R = \sqrt{(2At)}, \quad 0 < t < \infty;$$

$$k=-1: \quad R = \sqrt{(t^2 - A^2)}, \quad +A < t < \infty.$$

Show that in the case $k = 1$ the maximum 'radius' is A and the total lifetime is $2A$.

4. **The Einstein dust-filled universe** Show that including the cosmological term merely requires the left sides of (11.9) and (11.10) to be supplemented with the further term $-\Lambda$ in each case. Show that the adiabatic requirement (11.13) persists, and that therefore (11.14) still holds in a *dust-filled* universe.

The replacement for equation (11.15) is now

$$3(R'^2 + k)/R^2 - \Lambda = 3A/R^3.$$

This equation requires elliptic functions for its full solution. However, a qualitative treatment is possible along the lines of the treatment of orbits (section 9.7). Develop this idea.

5. Show that the coordinate change

$$t = T + (2H)^{-1}\ln(1 - H^2r^2), \quad \chi = r(1 - H^2r^2)^{-1/2}\exp - HT$$

transforms the Robertson–Walker metric (11.18) into the de Sitter metric (11.19).

6. Show that the coordinate change

$$\exp Ht_1 = (1 - H^2{\chi_2}^2 \exp - 2Ht_2)\exp Ht_2,$$
$$\chi_1 = \chi_2(1 - H^2{\chi_2}^2\exp - 2Ht_2)^{-1}\exp - 2Ht_2$$

transforms the expanding Robertson–Walker metric

$$dt_1{}^2 - \exp 2Ht_1\{d\chi_1{}^2 + \chi_1{}^2(d\theta^2 + \sin^2\theta d\phi^2)\}$$

into the contracting one

$$dt_2{}^2 - \exp - 2Ht_2\{d\chi_2{}^2 + \chi_2{}^2(d\theta^2 + \sin^2\theta\, d\phi^2)\}.$$

This is not at all mysterious, since the set of fundamental worldlines is quite different, though the backcloth is the same. The de Sitter spacetime can be 'cosmologized' in a variety of ways, for each value of k. A minor research project would be to find them all.

7. The cosmology with $R = t$ and $k = -1$, that is,

$$d\tau^2 = dt^2 - t^2(d\chi^2 + \sinh^2\chi(d\theta^2 + \sin^2\theta\, d\phi^2)),$$

is an *expanding* cosmology, since $R' > 0$. Show that there is a corresponding *contracting* cosmology, for which the metric takes the same form.

8. When $R = t$ and $k = -1$, every element of the Riemann tensor of (11.8) is zero everywhere; thus the metric of Problem 7 is flat. It is possible to have an expanding cosmology within a flat spacetime.

   Make this conclusion more explicit by making the coordinate change $T = t\cosh\chi$, $r = t\sinh\chi$, in the metric of Problem 7, thus reducing it to the Minkowski metric in the coordinates $Tr\theta\phi$ (see (4.3)). Sketch the $t\chi$-grid in the Minkowski diagram to get a feel for what is happening. (Note that this grid covers only one quadrant of the Minkowski diagram.)

   The situation here is not very different from the situation in Problems 5 and 6.

9. **An intermediate universe** The equations (11.15) and (11.17) are particular instances of the more general

$$R'^2 = -k + \frac{\alpha}{R} + \frac{\beta}{R^2},$$

   in which α and β are two positive constants. This equation controls the behaviour of a universe filled with material whose properties in some sense lie between the extremes of radiation and dust.

   Differentiate to obtain R'', and use (11.9) and (11.10) to find ρ and p. Hence show that the equation of state for the material is

$$\rho = 3p + (\text{constant}).\ p^{3/4}.$$

12

The interior equations for a spherically symmetric star

In the interior of a spherically symmetric mass distribution the energy–momentum–stress tensor is not zero, and the Schwarzschild metric does not apply. Finding an appropriate metric is important for two reasons: first, we now know that there exist stellar objects which are so small and so dense that Newtonian gravity is no longer good enough to describe their structure; second, General Relativity predicts that all sufficiently massive objects must inevitably collapse under their own weight. Since to be 'sufficiently massive' an object does not need to be much heavier than our Sun, gravitational collapse is of real practical interest to the astrophysicist.

This is a brief chapter; it is intended to give only a general overview.

12.1 The gravity field equations for a static spherical mass distribution

The general form of the appropriate static metric has been discussed in Chapter 8; it is (using geometrized units throughout, section 10.12)

$$ds^2 = A(r)\,dt^2 - B(r)\,dr^2 - r^2\,d\theta^2 - r^2\sin^2\theta\,d\phi^2, \tag{12.1}$$

involving the two disposable functions A and B. The Christoffel symbols and the Riemann tensor are given at (8.9) and (8.10); we shall need the tensor $\mathbf{G}$, with the four nonzero components

$$G^t{}_t \equiv -R^{r\theta}{}_{r\theta} - R^{r\phi}{}_{r\phi} - R^{\theta\phi}{}_{\theta\phi} = \frac{B'}{rB^2} + \frac{B-1}{r^2B},$$

$$G^r{}_r \equiv -R^{t\theta}{}_{t\theta} - R^{t\phi}{}_{t\phi} - R^{\theta\phi}{}_{\theta\phi} = -\frac{A'}{rAB} + \frac{B-1}{r^2B},$$

$$G^\theta{}_\theta \equiv -R^{tr}{}_{tr} - R^{t\phi}{}_{t\phi} - R^{r\phi}{}_{r\phi} = -\frac{A''}{2AB} + \frac{A'B'}{4AB^2} + \frac{A'^2}{4A^2B} - \frac{A'}{2rAB} + \frac{B'}{2rB^2},$$

$$G^\phi{}_\phi \equiv -R^{tr}{}_{tr} - R^{t\theta}{}_{t\theta} - R^{r\theta}{}_{r\theta} = G^\theta{}_\theta$$

(see Problem 1, Chapter 10). In empty space, these are all zero, leading to the familiar Schwarzschild solution. In regions occupied by matter,

$$G^t{}_t = 8\pi(\text{mass density}) \quad = 8\pi\rho, \tag{12.2}$$

$$G^r{}_r = 8\pi(\text{radial stress}) \quad = -8\pi P_1, \tag{12.3}$$

$$G^\theta{}_\theta = G^\phi{}_\phi = 8\pi(\text{transverse stress}) = -8\pi P_2. \tag{12.4}$$

It is usual for the radial and transverse stresses to be equal: $P_1 = P_2 = p$, the isotropic pressure. An important anisotropic case is considered in Problem 2.

When supplemented with an *equation of state* for the material of the star

$$\rho = \rho(p), \tag{12.5}$$

the equations are sufficient to yield the internal structure.

12.2 The field equations for the physical quantities

It is helpful to recast the field equations in a way which gives more prominence to the behaviour of the physical entities ρ and p: this will make the limits of validity of the static solution more evident.

The equation (12.2) may be rewritten

$$\frac{\mathrm{d}}{\mathrm{d}r}\left(r\frac{B-1}{B}\right) = 8\pi r^2 \rho. \tag{12.6}$$

This, with $B(0) = 1$ (required to ensure that there is no singularity at $r = 0$), gives

$$r\frac{B-1}{B} = \int_0^r 8\pi r^2 \rho \, \mathrm{d}r \equiv a(r), \text{ say,} \tag{12.7}$$

that is,

$$B(r) = \frac{1}{1 - a(r)/r}. \tag{12.8}$$

It will be convenient to use the function $a(r)$ in preference to $B(r)$. Apart from a factor 2, it has the appearance of being the total mass enclosed by the sphere of radius r. However, this interpretation is not quite correct; see Problem 1.

The equation (12.3) may now be written, using $a(r)$, as

$$-\frac{A'}{A}\left(1 - \frac{a(r)}{r}\right) + \frac{a(r)}{r^3} = -8\pi p,$$

or

$$\frac{A'}{A} = \frac{a(r) + 8\pi r^3 p}{r(r - a(r))}. \tag{12.9}$$

The equation (12.4) may also be rearranged, but an equivalent and more convenient procedure is to use one of the conservation equations which follow as a matter of identity (see Chapter 10, Problem 3). The equation in question is the r-component (in mixed form),

$$\partial_\mu T^\mu{}_r + \Gamma^\mu{}_{\mu\nu} T^\nu{}_r - \Gamma^\nu{}_{\mu r} T^\mu{}_\nu = 0,$$

summed, of course, over μ, $v = t, r, \theta, \phi$. There are 36 terms on the left; most are zero, however, since there are only 13 nonzero Γs, and only four nonzero components of **T**. The survivors are

$$\partial_r T^r{}_r + (\Gamma^t{}_{tr} + \Gamma^r{}_{rr} + \Gamma^\theta{}_{\theta r} + \Gamma^\phi{}_{\phi r})T^r{}_r - \Gamma^t{}_{tr}T^t{}_t - \Gamma^r{}_{rr}T^r{}_r - \Gamma^\theta{}_{\theta r}T^\theta{}_\theta - \Gamma^\phi{}_{\phi r}T^\phi{}_\phi = 0.$$

Two terms cancel; the remainder may be rewritten as

$$-\partial_r P_1 - \left(\frac{A'}{2A} + \frac{2}{r}\right)P_1 - \frac{A'}{2A}\rho + \frac{2}{r}P_2 = 0 \tag{12.10}$$

or, since $P_1 = P_2 = p$,

$$\frac{\mathrm{d}p}{\mathrm{d}r} + (p + \rho)\frac{A'}{2A} = 0. \tag{12.11}$$

Eliminating A'/A from (12.9) and (12.11) gives the **Oppenheimer–Volkov equation**,

$$\frac{\mathrm{d}p}{\mathrm{d}r} = -\frac{(p + \rho)(a(r) + 8\pi r^3 p)}{2r(r - a(r))}. \tag{12.12}$$

This is taken with the further equations

$$\frac{\mathrm{d}a(r)}{\mathrm{d}r} = 8\pi r^2 \rho, \quad a(0) = 0,$$

from (12.6), and

$$\rho = \rho(p) \quad \text{(the equation of state)}$$

to supply the recipe for the interior of a star, in terms of the physical density and pressure. Two coupled first-order differential equations require two disposable constants to specify the initial conditions. At $r = 0$, we must have $a(0) = 0$, leaving just one further available parameter; the pressure at the centre, $p(0)$, will do quite well.

Having chosen an equation of state, and a desired central pressure, the procedure is to integrate outwards from $r = 0$ *until the pressure drops to zero*; this is the signal that the surface of the star (r_S) has been reached. For $r > r_S$, both p and ρ are to be zero, and $a(r) \equiv a(r_S)$ is constant: outside the surface of the star the empty-space Schwarzschild solution applies.

12.3 How big can a star be?

Very few exact solutions using a realistic equation of state are known, and it is usual to explore the possibilities by integrating the differential equations numerically with the help of an electronic computer; this is a very straightforward process. However, to get a feel for the kind of thing that can happen, we now look at an exact solution which is on the edge of being realistic. This is the case of the *ultra-stiff* equation of state, $\rho =$ constant; the material of the star is incompressible.

Equation (12.7) gives immediately $a(r) = 8\pi r^3 \rho/3$, and the Oppenheimer–Volkov equation (12.12) becomes

$$\frac{\mathrm{d}p}{\mathrm{d}r} = -\frac{(p + \rho)(8\pi r^3 \rho/3 + 8\pi r^3 p)}{2r(r - 8\pi r^3 \rho/3)}.$$

This equation is separable:

$$\frac{dp}{(p+\rho)(3p+\rho)} = -\frac{4}{3}\pi\frac{r\,dr}{1-8\pi r^2\rho/3}.$$

Integrating gives

$$\left(\frac{3p+\rho}{p+\rho}\right)^2 = \left(\frac{3p_0+\rho}{p_0+\rho}\right)^2\left(1-\frac{8}{3}\pi r^2\rho\right), \tag{12.13}$$

where the constant of integration p_0 is the pressure at the centre of the star. To complete the solution, the function $A(r)$ is found by integrating (12.9) or (12.11); see Problem 4.

At the surface $r = r_S$ of the star, the pressure p is zero, and the left side of (12.13) is 1. Rearranging gives

$$r_S{}^2 = \frac{3}{2\pi\rho}\frac{p_0(2p_0+\rho)}{(3p_0+\rho)^2}. \tag{12.14}$$

This yields the radius of a star of uniform density ρ, where the pressure at the centre is p_0.

As one should expect, the larger the star, the greater the central pressure. However, (12.14) contains a surprise of momentous significance: *the star cannot exceed a certain critical size.* For, as $p_0 \to \infty$, the right side remains finite; in fact, r_S cannot exceed $R_{max} \equiv 1/\sqrt{(3\pi\rho)}$, however great the central pressure. The corresponding maximum mass M_{max} is $4R_{max}/9$. (This is a particular case of a general result, that all realistic equations of state lead to an upper limit on the size of a static star. An attempt to pack more mass into a volume than it can take will prohibit any stable configuration, and **gravitational collapse** will follow.)

This limit on the size of a star is easily reached. For example, it is believed that the density of a neutron star is roughly $10^{16}\,\mathrm{kg\,m^{-3}}$; the corresponding M_{max} is about 100 km, not all that much greater than the typical mass of the largest stars in the Galaxy. More reasonable equations of state lead to much smaller masses, not much greater than the mass of the Sun. Gravitational collapse cannot be an unusual phenomenon.

12.4 Gravitational collapse

There exists a simple model of gravitational collapse which shows all the main features to be expected. Envisage a large sphere occupied with a uniform dust; the pressure is to be zero everywhere and everywhen. In a sense, this model is utterly opposite to ultra-stiff; whatever the initial conditions, collapse is inevitable.

As it happens, we have met the relevant metrics already: outside of the sphere there is the expected Schwarzschild vacuum metric; inside there is a Robertson–Walker metric. These metrics are in contact along the worldtube swept out by the surface of the sphere. These assertions will now be justified for one case in particular.

Return to the Robertson–Walker metric for pressureless dust:

$$d\tau^2 = dt^2 - [R(t)]^2(d\chi^2 + \chi^2(d\theta^2 + \sin^2\theta\, d\phi^2))$$

where we have taken the case $k = 0$ for simplicity. The scale function is controlled by

(11.15) with $k = 0$:

$$R'^2 = A/R \ (A \text{ is constant}). \tag{12.15}$$

Since we are here considering, not an expansion, but a collapse, we shall have $R' < 0$.

The surface χ = constant encloses a spherical region of dust, which shrinks steadily as epoch passes. Without any loss of generality we take $\chi = 1$, absorbing any change into the scale factor R, which is now additionally the radius of the dust sphere. The $(1+2)$D worldtube swept out by this surface is defined by $d\chi = 0$, and its metric is therefore

$$dt^2 - R^2(d\theta^2 + \sin^2\theta\, d\phi^2). \tag{12.16}$$

Now turn to the Schwarzschild metric, and consider a spherical surface r = constant of observers in free radial fall. On the $(1+2)$D worldtube swept out by this shrinking sphere, use the coordinates θ, ϕ, and a new ω = the proper time common to all the observers; the metric will then be

$$d\omega^2 - r^2(d\theta^2 + \sin^2\theta\, d\phi^2). \tag{12.17}$$

The 'constant' r decreases as the sphere shrinks; its behaviour is controlled by (4.7a), and the case we need is $\gamma = 1$:

$$\left(\frac{dr}{d\omega}\right)^2 = \frac{a}{r}. \tag{12.18}$$

Compare the metrics (12.16) and (12.17), making the identifications $t{:}\omega$, $R(t){:}r(\omega)$, and $A{:}a$. Note that R and r are governed by the same equation (12.15) or (12.18), and may therefore be taken to be the same function. Thus *the two worldtubes are in every way identical.*

Hence we may do the following: (i) from the Robertson–Walker spacetime, keep only the worldtube and its dusty interior, and discard the rest; (ii) from the Schwarzschild spacetime, keep the worldtube and its empty exterior, and discard the rest; (iii) insert the result of (i) into the result of (ii). The foregoing discussion shows that the mutual fit will be perfect.

In this way, we obtain an exact picture of the history of gravitational collapse of a sphere of uniform dust. In this particular picture (with $k = 0$ and $\gamma = 1$) and dust sphere is originally ($t = -\infty$) very large, very sparse, and stationary; from there, it collapses to a black hole. Other values of k and γ are possible, and give different pictures.

12.5 The growth of the event horizon in gravitational collapse

Inside the dusty sphere it is convenient to parametrize the behaviour by

$$R = Ax^2, \quad t = 2Ax^3/3,$$

as $x \to 0$ through negative values; it is easy to check that (12.15) is satisfied. Equation (11.14) shows that $A/2$ is the total mass of the dust; since $R \to 0$, the dust must eventually fall within a black hole of this mass, that is, of radius $a = A$. Hence, *an event horizon of this radius must be formed.* In the remote past, there is no event horizon; the epoch of its first appearance, and the manner of its growth, are of interest.

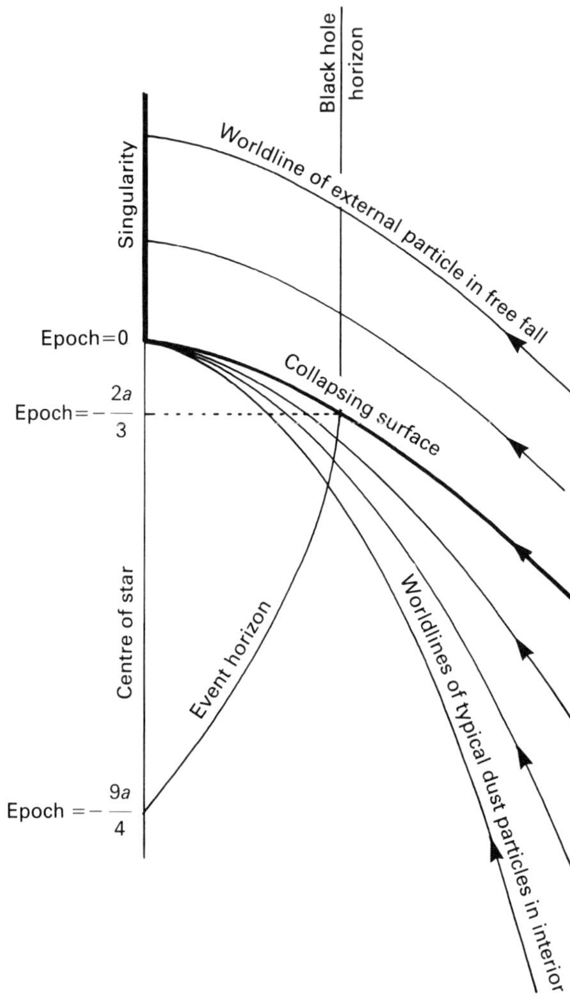

Fig. 12.1

The surface $r = a$ in the Schwarzschild metric intersects the worldtube of the dust at $R = a$; that is, at $x = -1$ and therefore at epoch $t = -2a/3$. Inside the worldtube, the event horizon will be defined by the radial null geodesic which arrives at the surface at this epoch. If this null-line starts from the centre $\chi = 0$ at the earlier epoch t_0, then—since the change in χ is from 0 to 1 (see section 11.4)—

$$1 = \int_{t_0}^{-2a/3} \frac{\mathrm{d}t}{R(t)} = \int_{x_0}^{-1} \frac{2ax^2\,\mathrm{d}x}{ax^2} = -2 - 2x_0.$$

Thus $x_0 = -3/2$, and the epoch $t_0 = -9a/4$. This is the epoch of first appearance of the event horizon.

12—Notes and problems

1. **Gravity gravitates** Equation (12.7) seems to suggest that $a(r)/2$ is a straightforward integral of the density ρ, and is thus the total energy inside the sphere of radius r. This interpretation would be wrong. The 'correct' volume element for integration in the metric in question is not the simple 'flat' $\mathrm{d}r \cdot r\,\mathrm{d}\theta \cdot r \sin\theta\,\mathrm{d}\phi$, but rather $\sqrt{(-\det|g|)}\,\mathrm{d}r\,\mathrm{d}\theta\,\mathrm{d}\phi = \sqrt{B}\,\mathrm{d}r \cdot r\,\mathrm{d}\theta \cdot r \sin\theta\,\mathrm{d}\phi$. It is therefore necessary to rewrite the integral as

$$\frac{a(r)}{2} = \int_0^r 4\pi r^2 \rho \sqrt{\left(1 - \frac{a(r)}{r}\right)} \frac{\mathrm{d}r}{\sqrt{(1 - a(r)/r)}}.$$

Thus the 'true' density, naively, is

$$\rho \sqrt{\left(1 - \frac{a(r)}{r}\right)} = \rho - \frac{\rho\, a(r)/2}{r} + \cdots.$$

The second term on the right is the Newtonian potential energy of the density ρ in the presence of the mass $a(r)/2$ inside the sphere of radius r. Integrating over all space gives: total mass + total Newtonian potential energy + further correction terms. In estimating the gravitational effects of a body on its environment, *all* kinds of mass or energy must be taken into account, *even gravitational energy*. This is the physical consequence of the mathematical fact that the field equations are nonlinear.

2. The methods of this chapter allow us to deal with the problem of a **spherical charged mass**. The environment of a static electric charge is not a vacuum, since it is filled with a static electric field, with its own energy–momentum–stress tensor. In fact, associated with the field E at any point is an energy density $E^2/2$, a tension $E^2/2$ in the direction of the field, and a pressure $E^2/2$ in the perpendicular directions. Thus for a spherical charged mass, for which the electric field must be radial, we can assert that $\rho = -P_1 = P_2$ in (12.2)–(12.4).

   Finding the appropriate metric is quite simple in this case. Show the following:

   (i) In (12.2), (12.3), $G^t{}_t - G^r{}_r = 0$, leading to $AB = \text{constant}$.

   (ii) Equation (12.10) collapses to $\partial_r \rho + (4/r)\rho = 0$, with solution $\rho = b/8\pi r^4$, where b is a disposable constant. (This is effectively the inverse-square law for the electric intensity E.)

   (iii) Equation (12.6) is still valid, and yields—with the ρ just found—

   $$B = \frac{1}{1 - \dfrac{a}{r} + \dfrac{b}{r^2}},$$

   where a is a second disposable constant.

   Hence the metric for this problem is

$$\mathrm{d}s^2 = (1 - a/r + b/r^2)\,\mathrm{d}t^2 - \frac{\mathrm{d}r^2}{1 - a/r + b/r^2} - r^2(\mathrm{d}\theta^2 + \sin^2\theta\,\mathrm{d}\phi^2).$$

As suggested in Problem 1, the constant $a/2$ ought to be regarded as the total of all kinds of identifiable energy, of whatever origin: the mass of the sphere itself, the total energy of the electric field, the gravitational self-interaction of this field, etc.

The constant b is more or less the square of the charge on the sphere, in geometrized units. Show that if $b > a^2/4$ the metric has no singularities except at $r = 0$, there is no event horizon, and the entire spacetime is covered by the coordinates $tr\theta\phi$.

The metric for a spherically symmetric charged mass was first discussed by Reissner in 1916.

3. The maximum possible radius of a star of given constant density is (9/8) of the radius of a black hole of the same mass.

4. In section 12.3 the problem for constant density is only partly solved. To obtain the metric, first note that (12.7) with $a(r) = 8\pi r^3 \rho/3$ gives

$$\text{for } r < r_S, \quad B(r) = \frac{1}{1 - 8\pi r^2 \rho/3},$$

$$\text{for } r > r_S, \quad B(r) = \frac{1}{1 - 8\pi {r_S}^3 \rho/3r},$$

the second being the obligatory Schwarzschild form to be expected outside the surface of the sphere. The boundary condition that $B(r)$ is to be continuous is satisfied at $r = r_S$.

As for $A(r)$, the equation (12.11) is very simple when $\rho =$ constant; it integrates without difficulty to

$$\sqrt{A} = \frac{\text{constant}}{p + \rho}.$$

Show from (12.13) that

$$\frac{2\rho}{p + \rho} = 3 - \frac{\sqrt{(1 - 8\pi r^2 \rho/3)}}{\sqrt{(1 - 8\pi {r_S}^2 \rho/3)}},$$

and hence that

$$\text{for } r < r_S, \quad \sqrt{A} = \frac{3}{2}\sqrt{\left(1 - \frac{8\pi {r_S}^2 \rho}{3}\right)} - \frac{1}{2}\sqrt{\left(1 - \frac{8\pi r^2 \rho}{3}\right)},$$

$$\text{for } r > r_S, \quad A = 1 - \frac{8\pi {r_S}^3 \rho}{3r},$$

the second being again the obligatory Schwarzschild form.

13

Weak gravity and gravity waves

In the last few chapters we have been looking at gravity in circumstances where it may be expected to have an enormous influence on the course of events—gravitational collapse, cosmology, and so on. Here we shall return to 'everyday' gravity, which is so weak as to allow for all calculations to be done in the *linear approximation*. Well-known techniques, involving the wave-equation in particular, become available, and the whole theory becomes much simpler.

In Chapter 4 we used a low-order approximation for geodesics in the Schwarzschild solution; now we are concerned with a low-order approximation for the solutions of the Einstein gravity field equation. Apart from the Lense–Thirring effect (section 13.13), this chapter is concerned almost exclusively with gravitational waves.

13.1 The weak-field approximation

The **linear approximation** (or **weak-field approximation**) assumes that the metric differs only slightly from the Minkowski form:

$$g_{\mu\nu} = \gamma_{\mu\nu} + h_{\mu\nu} \tag{13.1}$$

where

$$\gamma_{\mu\nu} = \mathrm{diag}(1, -1, -1, -1) \tag{13.2}$$

and $h_{\mu\nu}$ is small enough for terms with more than one factor h to be discarded. (We shall use geometrized units throughout; see section 10.12.) In this approximation it is usually possible to use $\boldsymbol{\gamma}$, rather than $\mathbf{g}$, to raise and lower affixes, since the important tensors are of the same order of smallness as $\mathbf{h}$. The only notable exception is $\mathbf{g}$ itself, for which

$$g_{\mu\nu} = \gamma_{\mu\nu} + h_{\mu\nu}; \quad g^{\mu}{}_{\nu} = \gamma^{\mu}{}_{\nu} (= \delta^{\mu}{}_{\nu}, \text{ of course}); \quad g^{\mu\nu} = \gamma^{\mu\nu} - h^{\mu\nu}.$$

(See Problem 1.)

Within this approximation, the definition of Γ of (5.2) becomes

$$\Gamma^{\mu}{}_{\alpha\beta} = \tfrac{1}{2}\gamma^{\mu\nu}(\partial_\alpha h_{\beta\nu} + \partial_\beta h_{\alpha\nu} - \partial_\nu h_{\alpha\beta}), \tag{13.3}$$

in which it has been possible to make one replacement of $\mathbf{g}$ by γ. In the definition (6.8) of the Riemann tensor, the terms quadratic in the Γs are negligibly small, and there remains

$$R^{\mu}{}_{\nu\rho\sigma} = \partial_\rho \Gamma^{\mu}{}_{\sigma\nu} - \partial_\sigma \Gamma^{\mu}{}_{\rho\nu}. \tag{13.4}$$

Inserting (13.3) and lowering the affix μ yields (after a slight cancellation)

$$R_{\mu\nu\rho\sigma} = \tfrac{1}{2}(-\partial_\mu\partial_\rho h_{\nu\sigma} - \partial_\nu\partial_\sigma h_{\mu\rho} + \partial_\mu\partial_\sigma h_{\nu\rho} + \partial_\nu\partial_\rho h_{\mu\sigma}). \tag{13.5}$$

It is easy to confirm that this expression displays all the symmetries required by the work of section 6.4.

13.2 Plane gravity waves in empty space

The possibility that gravity oscillations may be emitted by cataclysmic events in the Universe has attracted considerable attention in recent years. Theoretically, Einstein's linearized field equations possess wave-like solutions; to begin with we look at plane waves in empty space ($T_{\mu\nu} = 0$) in the linear approximation.

A **plane gravity wave** is a small periodic perturbation of the metric of spacetime of the form

$$h_{\mu\nu} = \mathrm{Re}\,\{a_{\mu\nu} \exp ik_\mu x^\mu\}. \tag{13.6}$$

(Re $\{\ldots\}$ stands for *real part of*.) This involves the **propagation vector**

$$k_\mu = (\omega, -\mathbf{k}), \quad k_\mu x^\mu = \omega t - \mathbf{k}\cdot\mathbf{r}.$$

The amplitudes $a_{\mu\nu}$ are constants, possibly complex to allow for a variety of phase relationships between the components. Differentiating twice yields

$$\partial_\rho \partial_\sigma h_{\mu\nu} = -k_\rho k_\sigma h_{\mu\nu}. \tag{13.7}$$

Thus the Riemann tensor in this case is

$$R_{\mu\nu\rho\sigma} = \tfrac{1}{2}(k_\mu k_\rho h_{\nu\sigma} + k_\nu k_\sigma h_{\mu\rho} - k_\mu k_\sigma h_{\nu\rho} - k_\nu k_\rho h_{\mu\sigma}). \tag{13.8}$$

The Ricci tensor follows immediately:

$$\begin{aligned} R_{\nu\sigma} &= \gamma^{\mu\rho} R_{\mu\nu\rho\sigma} \qquad \text{(to sufficient accuracy)} \\ &= \tfrac{1}{2}(k^2 h_{\nu\sigma} - k_\nu w_\sigma - k_\sigma w_\nu), \end{aligned} \tag{13.9}$$

where for convenience we have written

$$k^2 \text{ for } k_\mu k^\mu$$

and

$$w_\nu \text{ for } h^{\alpha}{}_{\nu} k_\alpha - \tfrac{1}{2} h^{\rho}{}_{\rho} k_\nu.$$

The empty-space field equations Ricci $= 0$ therefore imply

$$k^2 h_{\nu\sigma} = k_\nu w_\sigma + k_\sigma w_\nu.$$

There are two cases.

$\mathbf{k}^2 \neq 0$ Here there is no interest, since $h_{\nu\sigma}$ is very restricted in form,

$$h_{\nu\sigma} = k^{-2}(k_\nu w_\sigma + k_\sigma w_\nu).$$

Substituting in (13.8) leads immediately to $R_{\mu\nu\rho\sigma} = 0$ everywhere. So *the spacetime is flat*. The wave is not a physical wave at all, but merely a periodic flexure of the coordinate system.

$\mathbf{k}^2 = 0$ Here a *physical* wave becomes possible; such a wave propagates—since **k** is null—with the speed of light. (It is perhaps more appropriate to say that light travels at the speed of gravity!) The field equations now give

$$k_\nu w_\sigma + k_\sigma w_\nu = 0,$$

easily shown to imply $w_\sigma = 0$, that is

$$h^\rho{}_\sigma k_\rho = h^\alpha{}_\alpha k_\sigma / 2. \tag{13.10}$$

This requirement is the **gauge condition**; it is not strong enough to make the curvature vanish, though it does lead to the less stringent restriction

$$R_{\mu\nu\rho\sigma} k^\sigma = 0. \tag{13.11}$$

The tensor $h_{\mu\nu}$ is symmetric, and thus has ten independent components. The gauge condition places four conditions on these components, reducing the freedom to *six*. The possibility of small arbitrary adjustments to the four coordinates accounts for a fourfold freedom which neither changes the *physical* situation, nor affects the validity of the gauge condition (it is after all a *tensor* requirement). We conclude that the remaining twofold freedom is all that there is with physical significance, and we say that a plane wave with propagation vector **k** may be represented as the superposition of just two **polarizations**.

13.3 How a plane wave manifests itself

No single observer can be aware of the entire Riemann tensor, but only that part of it which is contained in the appropriate tidal tensor

$$\Delta_{\mu\sigma} = R_{\mu\nu\rho\sigma} \dot{x}^\nu \dot{x}^\rho$$

in which $\dot{x}^\nu$ is the observer's four-velocity. As always, $\Delta_{\mu\sigma}$ is *symmetric, spacelike and traceless*:

$$\Delta_{\mu\sigma} = \Delta_{\sigma\mu}, \quad \Delta_{\mu\sigma}\dot{x}^\sigma = 0, \quad \Delta^\mu{}_\mu = 0. \tag{13.12}$$

(We have seen that the first two are always true, by the symmetries of the Riemann tensor. The third is expected to be true in empty spacetime; indeed, we used it in section 8.2 to obtain the appropriate field equations.) Additionally, for a plane wave, the tidal tensor is *transverse*:

$$\Delta_{\mu\sigma} k^\sigma = 0; \tag{13.13}$$

this follows from (13.11).

To fix ideas, consider a stationary observer and a plane wave propagating in the

z-direction:

$$x^\mu = (t, x, y, z),$$
$$\dot{x}^\mu = (1, 0, 0, 0),$$
$$k^\mu = (\omega, 0, 0, \omega).$$

In this case the most general $\Delta_{\mu\nu}$ which satisfies the requirements is

$$\mathbf{\Delta} = \mathrm{Re}\left\{\begin{bmatrix} 0 & 0 & 0 & 0 \\ 0 & a & b & 0 \\ 0 & b & -a & 0 \\ 0 & 0 & 0 & 0 \end{bmatrix} \exp i\omega(t - z)\right\} \tag{13.14}$$

where a and b are two disposable, possibly complex, constants. It follows that all possible polarizations of the wave may be obtained by superposing just two, with arbitrary amplitudes and arbitrary relative phase. We shall consider two examples.

Plane polarization Two neighbouring test particles in the plane $z = 0$ are separated by a vector

$$\boldsymbol{\lambda} = (\lambda_x, \lambda_y, 0).$$

Set $a = 1$, $b = 0$. The mutual tidal acceleration of the test particles is

$$\mathrm{Re}\left\{\begin{bmatrix} 1 & 0 & 0 \\ 0 & -1 & 0 \\ 0 & 0 & 0 \end{bmatrix}\begin{bmatrix} \lambda_x \\ \lambda_y \\ 0 \end{bmatrix} \exp i\omega t\right\} = \begin{bmatrix} \lambda_x \\ -\lambda_y \\ 0 \end{bmatrix} \cos \omega t.$$

Integrating twice gives the mutual *true* displacement from 'equilibrium'

$$\mathbf{d} = -\omega^{-2}(\lambda_x, -\lambda_y, 0) \cos \omega t$$

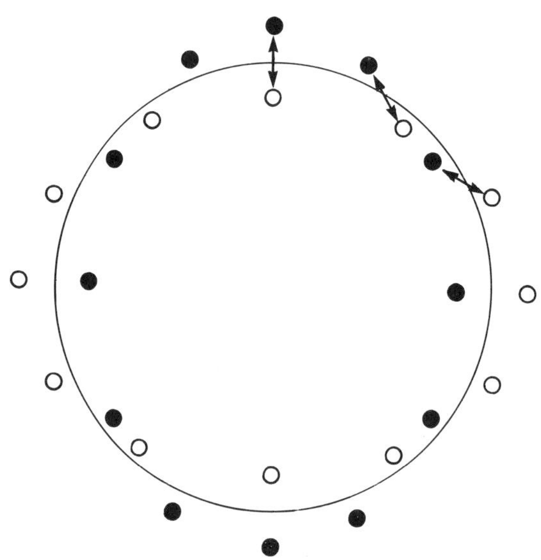

Fig. 13.1

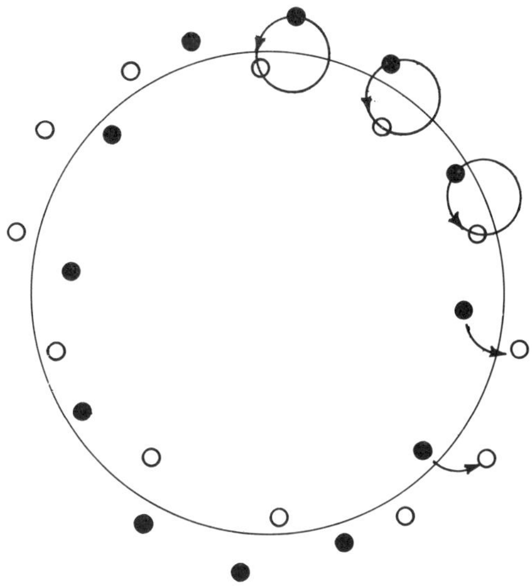

Fig. 13.2

(ignoring constants of integration, one of which gives a 'drift' term). The particles execute simple harmonic motion with respect to each other. Fig. 13.1 gives the relation between **d** and **λ** at $t = 0$, and shows the effect on a ring of free particles as time passes.

Other *real* values of a and b lead to similar results: these are all plane polarizations, differing only in the orientation of the pattern.

Circular polarizations Setting $a = 1$, $b = \pm \mathrm{i}$ gives the two possible circular polarizations. With the + sign, we can work out

$$\mathbf{d} = -\omega^{-2}(\lambda_x \cos \omega t - \lambda_y \sin \omega t, \ -\lambda_x \sin \omega t - \lambda_y \cos \omega t, 0)$$

with the effect of a rotation as in Fig. 13.2. The opposite choice of sign gives a rotation in the opposite sense.

Clearly many other types of superposition are possible.

Two ways of detecting gravity waves have been suggested. One is to install a family of satellites in free fall in orbit, and continuously to monitor their mutual separations, by radar methods, say. Gravitational disturbances will then appear as discrepancies between the observed and the calculated relative orbits. The other, older, method is to search for periodic deformations in an 'antenna'—a delicately suspended solid object such as an aluminium cylinder with attached piezoelectric transducers. Weber has claimed to have detected gravity radiation by the older method, but other workers have so far failed to reproduce his results.

13.4 The general gravitational disturbance in empty spacetime

Since

$$\partial_\sigma \exp \mathrm{i}(k_\mu x^\mu) = \mathrm{i}k_\sigma \exp \mathrm{i}(k_\mu x^\mu),$$

much of the foregoing may be written in a differential form. For example, the condition for a physical gravity wave

$$h^{\rho}{}_{\sigma}k_{\rho} - \tfrac{1}{2}h^{\alpha}{}_{\alpha}k_{\sigma} = 0$$

becomes the **gauge condition** (Problem 2)

$$\partial_{\rho}H^{\rho}{}_{\sigma} = 0,$$

in which we have introduced the alternative and in many ways more useful expression of the wave

$$H^{\rho}{}_{\sigma} \equiv h^{\rho}{}_{\sigma} - \tfrac{1}{2}\gamma^{\rho}{}_{\sigma}h^{\alpha}{}_{\alpha}.$$

The requirement

$$k_{\sigma}k^{\sigma} = 0 \quad \text{becomes the } \textbf{wave equation} \quad \Box H_{\mu\nu} = 0.$$

where the **d'Alembertian operator** (often pronounced '*square*') is defined as

$$\Box \equiv \gamma^{\alpha\beta}\partial_{\alpha}\partial_{\beta} = \frac{\partial^2}{\partial t^2} - \nabla^2.$$

Any general linear superposition of such physical waves will also satisfy these differential equations, whilst any 'wave' which is merely a coordinate deformation may safely be left out. Indeed, by the Fourier theorem, there are enough such waves to allow us to build up *any* solution of the differential equations.

We conclude that the equations for a general gravitational disturbance in empty spacetime may be taken to be

$$\Box H_{\mu\nu} = 0 \quad \text{and} \quad \partial_{\mu}H^{\mu}{}_{\nu} = 0;$$

thus, the wave equation and the gauge condition may be taken to be separately satisfied.

13.5 The equations with a source term

We must now consider the generation of gravity waves by matter. This is governed by the Einstein field-equation with a nonzero *source* (10.16). The Einstein tensor may be calculated from (13.5) in the usual way (see Problem 10.1); in terms of the **H** introduced above it becomes

$$8\pi T_{\mu\nu} = G_{\mu\nu} = -\tfrac{1}{2}\Box H_{\mu\nu} + \tfrac{1}{2}\partial_{\mu}\partial_{\alpha}H^{\alpha}{}_{\nu} + \tfrac{1}{2}\partial_{\nu}\partial_{\alpha}H^{\alpha}{}_{\mu} - \tfrac{1}{2}\gamma_{\mu\nu}\partial_{\alpha}\partial_{\beta}H^{\alpha\beta}. \tag{13.15}$$

Thus $T_{\mu\nu}$—which represents the distribution of energy, momentum and stress—appears as a source term in a differential equation for $H_{\mu\nu}$. It turns out that it is always possible to adjust the coordinate system so that the equations

$$\underset{\text{(wave equation)}}{8\pi T_{\mu\nu} = -\tfrac{1}{2}\Box H_{\mu\nu}} \quad \text{and} \quad \underset{\text{(gauge condition)}}{\partial_{\alpha}H^{\alpha}{}_{\nu} = 0}$$

are *separately* satisfied. This is entirely in line with the result of the last section, and often leads to useful simplifications. (The demand that the equations must be separately satisfied effectively excludes the uninteresting solutions for which k^2 is not zero.) In this formulation, the inevitable conservation law

$$\partial_\alpha T^\alpha{}_\nu = 0$$

follows as a consequence of the gauge condition.

The wave-equation is easily solved with the help of any **Green function** $D(t, \mathbf{r})$ satisfying

$$\Box D(txyz) = \delta(t)\delta(x)\delta(y)\delta(z)$$

where $\delta(x)$ is the Dirac-function (see Problem 3). For then the function

$$H_{\mu\nu} = -16\pi \int_{\text{all spacetime}} \mathrm{d}t'\, \mathrm{d}^3\mathbf{r}'\, D(t-t', \mathbf{r}-\mathbf{r}')T_{\mu\nu}(t', \mathbf{r}')$$

satisfies the equation

$$\begin{aligned}\Box H_{\mu\nu} &= -16\pi \int \mathrm{d}t'\, \mathrm{d}^3\mathbf{r}'\delta(t-t')\delta(x-x')\delta(y-y')\delta(z-z')T_{\mu\nu}(t'x'y'z') \\ &= -16\pi T_{\mu\nu}(t, \mathbf{r}),\end{aligned}$$

as required by the usual property of the δ-function.

There are many possibilities for the Green function D: they differ from each other by some solution of the wave-equation without the source term. Almost always the most appropriate one in physical applications is the **retarded propagator**,

$$D_{\mathrm{R}}(t, \mathbf{r}) = \begin{cases} \delta(t^2 - \mathbf{r}^2)/2\pi & (t > 0) \\ 0 & (t < 0) \end{cases}.$$

It is manifestly Lorentz invariant, and *is zero except on the future null cone.* Thus the effects of a nonzero $\mathbf{T}$ propagate outwards as future time passes, at the speed of light. Additionally for this particular choice of Green's function, the gauge condition is normally satisfied, and is certainly satisfied when $\mathbf{T}$ is nonzero in a worldtube of finite cross-section.

By a property of the δ-function, D_{R} may be written

$$D_{\mathrm{R}}(t, \mathbf{r}) = \frac{\delta(t-r)}{4\pi r}, \quad (r \equiv |\mathbf{r}|)$$

and therefore

$$H_{\mu\nu}(t, \mathbf{r}) = -4 \int_{\text{all space}} \mathrm{d}^3\mathbf{r}' \frac{T_{\mu\nu}(t - |\mathbf{r}-\mathbf{r}'|, \mathbf{r}')}{|\mathbf{r}-\mathbf{r}'|}. \tag{13.16}$$

Explicitly, $\mathbf{H}$ at the position $\mathbf{r}$ is represented as a superposition of contributions from $\mathbf{T}$ at other positions $\mathbf{r}'$ at the **retarded times** $t - |\mathbf{r}-\mathbf{r}'|$.

13.6 The compact source approximation

An important simplification occurs where the region of nonzero $\mathbf{T}$ occupies a very narrow worldtube, so narrow that the dimensions of its cross-section may be ignored. (What we have in mind is the case of a *small object, compact* in the sense of section 10.8.) By momentum conservation, the worldtube will be straight, and it is usually convenient to arrange for it to lie along the t-axis. To sufficient approximation we may then set—since $\mathbf{r}'$ is negligible—

$$|\mathbf{r}-\mathbf{r}'|=|\mathbf{r}|=r \text{ say,}$$

and the integration of (13.16) collapses to

$$H_{\mu\nu}(t,\mathbf{r})=\frac{1}{r}K_{\mu\nu}(t-r) \tag{13.17}$$

where

$$K_{\mu\nu}(t-r)=-4\int_{\text{cross-section}} \mathrm{d}^3\mathbf{r}'\,T_{\mu\nu}(t-r,\mathbf{r}'). \tag{13.18}$$

The work has been reduced to integrating $\mathbf{T}$ across the tube at the *fixed retarded time* $t-r$. This approach concentrates attention on that part of the field which goes like r^{-1}; this is known as the **far field**. Correction terms fall off more rapidly, and at large distances become negligible in comparison. (The Lense–Thirring effect involves such a correction term; see section 13.13.)

An elementary example will show how things go. Consider the gravity field of *stationary* matter distributed in the neighbourhood of $\mathbf{r}'=0$ with density $\rho(\mathbf{r}')$; in this case

$$T_{tt}=\rho,$$

other components being zero, or at least negligible. The obvious integration shows that the only nonzero component of $\mathbf{K}$ is

$$K_{tt}=-4M$$

where M is the total integrated mass. (13.17) immediately gives the only nonzero component of $\mathbf{H}$:

$$H_{tt}=\frac{-4M}{r}.$$

To obtain the resulting metric we need $h_{\mu\nu}$, rather than $H_{\mu\nu}$, from

$$h_{\mu\nu}\equiv H_{\mu\nu}-\tfrac{1}{2}\gamma_{\mu\nu}H^{\alpha}{}_{\alpha};$$

see Problem 4. In the present case, only the diagonal terms are nonzero; they are

$$h_{tt}=h_{xx}=h_{yy}=h_{zz}=\frac{-2M}{r}.$$

Taking this with

$$\gamma_{tt}=1,\quad \gamma_{xx}=\gamma_{yy}=\gamma_{zz}=-1$$

leads to the metric

$$\mathrm{d}s^2=(1-2M/r)\,\mathrm{d}t^2-(1+2M/r)(\mathrm{d}x^2+\mathrm{d}y^2+\mathrm{d}z^2).$$

This is in fact a first-order approximation to the Schwarzschild metric, in slightly unfamiliar coordinates.

Clearly the compact source approximation is best when distances and wavelengths are large compared with the size of the source. In the example just given, r is to be large compared with $2M$: thus the Schwarzschild radius is a measure of how compact a source may be taken to be. From this point of view, all the bodies in the Solar System are extremely compact—the Sun is the least compact, with an M of the order of a kilometre—and the linear approximation may be expected to be very good. (The approximations made in the treatment of the evidences for General Relativity within the Solar System are of this order of magnitude; see Chapter 4.)

13.7 The inertia tensor

We are interested in the more general case where other components of **T** may be nonzero, and even time-dependent. We know from the work of section 10.8 that the integrated mass of the source must be constant, and that the worldtube of the source must be straight, in a direction determined by the (constant) total momentum. Without any loss of generality, we take this direction to be the t-axis: the momentum is then zero. Thus generally

$$K_{tt} = -4M \text{ (constant)}, \quad K_{tx} = K_{ty} = K_{tz} = 0 = K_{xt} = K_{yt} = K_{zt}.$$

The remaining nine components are effectively the integrated stress within the source,

$$K_{ij} = -4\int t_{ij}\,\mathrm{d}^3\mathbf{r}. \tag{13.19}$$

Thus, in the compact source approximation, the far field of the source falls into two parts:

(a) a steady field from the total constant mass M, and
(b) a possibly varying field arising from the integrated internal stresses.

More detailed information is irrelevant for the far field and for long wavelengths. Clearly, it is (b) which will be responsible for any emitted radiation.

It is usually inconvenient to evaluate the integrated internal stress directly. For a compact source there is an alternative route which depends on a relationship of the kind already considered in section 10.8. Here we take the divergence

$$\begin{aligned}\partial_i(p^i r^j r^k) &= r^j r^k \nabla\cdot\mathbf{p} + p^i\,(\partial_i r^j) r^k + p^i r^j(\partial_i r^k)\\ &= r^j r^k \nabla\cdot\mathbf{p} + p^j r^k + p^k r^j\end{aligned}$$

and integrate over any volume whose boundary encloses the compact source entirely.

By the Gauss Theorem, the left side gives zero. The first term on the right gives (see section 10.8a)

$$\int_V r^j r^k \nabla\cdot\mathbf{p}\,\mathrm{d}^3\mathbf{r} = -\frac{\mathrm{d}}{\mathrm{d}t}\int_V r^j r^k \rho\,\mathrm{d}^3\mathbf{r} \equiv -\frac{\mathrm{d}}{\mathrm{d}t}\mu^{ik}$$

where μ^{ij} is the **inertia tensor** of the compact source. Its elements are the moments and products of inertia of the compact source at its centre of mass. They are easily found if the mass distribution ρ is known. As for the second and third terms, the work of section 10.8d shows that the time derivative of either is the integrated stress.

Hence

$$\frac{d^2}{dt^2}\mu^{jk} = 2\int t^{jk}\,d^3\mathbf{r}; \tag{13.20}$$

this shows how the integrated stress may be conveniently found by way of the inertia tensor when we are dealing with a compact source.

The far field which results from a varying inertia tensor is given by

$$H^{tt} = 0; \quad H^{ti} = 0 \quad (i = x, y, z); \quad H^{jk}(t, \mathbf{r}) = -2r^{-1}\frac{d^2\mu^{jk}(t-r)}{dt^2} \tag{13.21}$$

by equations (13.17), (13.19) and (13.20).

13.8 The traceless-transverse gauge

It is useful to be able to extract from the tensor **H** that part which has a physical effect. An observer at a fixed point (that is, with the four-velocity $(1, 0, 0, 0)$) experiences a gravity wave with a definite frequency ω and definite direction of propagation $\mathbf{u}$; the propagation vector is of course $k_\mu = (\omega, \omega\mathbf{u})$. At that point the disturbance is specified by the tensor $H_{\mu\nu}$.

The physical effects of the wave, as far as this observer is concerned, are enshrined in the tidal tensor, as defined in section 8.1. This tidal tensor contains, among many other things, a term

$$\Delta_{\mu\nu} = -\tfrac{1}{2}\omega^2 H_{\mu\nu} + \cdots$$

arising from the last term in (13.8). The entire expression must satisfy (13.12), while the first term may not; it is a useful fact that this enables us to fill out the rest of the expression without much trouble.

The four-velocity of the observer is $(1, 0, 0, 0)$, by which we infer, using (13.12),

$$\Delta_{tt} = 0, \quad \Delta_{ti} = \Delta_{it} = 0, \quad \text{for } i = x, y, z. \tag{13.22}$$

Thus $\boldsymbol{\Delta}$ is effectively a 3D tensor, and we shall concentrate on the nine components Δ_{ij} for i and $j = x$ or y or z. On the other hand, the tensor **H** of interest to us is also effectively 3D, for a different reason: it arises as the far field generated by a varying inertia tensor; see equation (13.21).

The trick is to assemble all the 3D tensor expressions which involve **H** and **u** in a sensible way. They are

$$H_{ij} \text{ itself}, \ \delta_{ij}, \ u_i u_j, \text{ and } u_i u_k H_{kj}.$$

(Note that we shall write all spacelike affixes as suffixes: there is no point in distinguishing. Moreover, in $u_k H_{kj}$ the summation convention is to apply to the repeated suffix k.) We attempt to define the **traceless-transverse (TT)** version of **H** by writing

$$H_{ij}^{\mathrm{TT}} = H_{ij} + A\delta_{ij} + Bu_i u_j + C[u_i u_k H_{kj} + u_j u_k H_{ki}] \tag{13.23a}$$

and choosing A, B, C in such a way that H_{ii}^{TT} and $H_{ij}^{\mathrm{TT}} u_j$ are both zero. This can be done in exactly one way, and

$$2A = u_i H_{ij} u_j - H_{ii}; \quad 2B = u_i H_{ij} u_j + H_{ii}; \quad C = -1. \tag{13.23b}$$

The recipe for obtaining $\mathbf{\Delta}$ from $\mathbf{H}$ is to evaluate $\mathbf{H}^{\mathrm{TT}}$ according to these rules, and then to write

$$\Delta_{\mu\nu} = -\tfrac{1}{2}\omega^2 H_{\mu\nu}^{\mathrm{TT}}. \tag{13.24}$$

One may suspect that the tensors $\mathbf{H}$ and $\mathbf{H}^{\mathrm{TT}}$ are in some sense equivalent, since they seem to contain the same physical information. In fact, the fourfold coordinate freedom permits either to be transformed into the other without any change of physical content. We may use either according to convenience. To describe the global radiation from a changing inertia tensor $\mathbf{H}$ is best. To describe the effect of such radiation on a remote stationary observer, change to $\mathbf{H}^{\mathrm{TT}}$ using (13.23).

13.9 Radiation from an oscillating inertia tensor

The tidal tensor for an observer at location $r\mathbf{u}$ arising from radiation by an oscillating term $\mu_0 \cos \omega t$ in the inertia tensor located at the origin is, by (13.24) and (13.21),

$$\Delta_{ij} = [-\omega^4 r^{-1} \mu_{0ij} \cos \omega(t-r)]^{\mathrm{TT}} \tag{13.25}$$

where the direction of propagation implied in the TT is radially outward along $\mathbf{u}$. The remaining elements are zero; see (13.22). Thus, apart from the $1/r$ dependence and the phase change due to the finite speed of propagation, what the observer 'feels' from the source is not the full inertia tensor, but only that part which is traceless, and transverse to his line of sight. This gives a very simple prescription for finding the polarization of the radiation in different directions.

For example, take $\mu_{zz} = \mu_0 \cos \omega t$, the other elements being zero. For an observer on the x-axis, $\mathbf{u} = (1, 0, 0)$. We need the TT-version of the matrix

$$\mathbf{M} = \begin{bmatrix} 0 & 0 & 0 \\ 0 & 0 & 0 \\ 0 & 0 & \mu_0 \end{bmatrix}.$$

The prescription (13.23b) gives $A = -B = \mu_0/2$; the third term of (13.23a) is in fact zero; hence

$$\mathbf{M}^{\mathrm{TT}} = \begin{bmatrix} 0 & 0 & 0 \\ 0 & 0 & 0 \\ 0 & 0 & \mu_0 \end{bmatrix} - \tfrac{1}{2}\mu_0 \begin{bmatrix} 1 & 0 & 0 \\ 0 & 1 & 0 \\ 0 & 0 & 1 \end{bmatrix} + \tfrac{1}{2}\mu_0 \begin{bmatrix} 1 & 0 & 0 \\ 0 & 0 & 0 \\ 0 & 0 & 0 \end{bmatrix}$$

$$= \tfrac{1}{2}\mu_0 \begin{bmatrix} 0 & 0 & 0 \\ 0 & -1 & 0 \\ 0 & 0 & 1 \end{bmatrix}.$$

According to the discussion of section 13.3, this represents a plane polarization in the yz-plane.

For an observer on the z-axis, $\mathbf{M}^{\mathrm{TT}}$ turns out to be zero: there is no radiation in the z-direction. The nature of the radiation in more general directions may be obtained similarly.

13.10 Does a gravity wave carry energy?

No, it doesn't.

There is no reason why it should. Since a gravity wave is a weak *distortion* of Minkowski space, the Lorentz symmetry properties (which guarantee—among other things—the conservation of energy) are lost. In particular, an absorber may increase its energy as a result of the effect of a gravity wave, even though the wave itself has no energy to impart to it. This is entirely in accord with the field equations of empty space $\mathbf{G}=0$, three of which say that the energy current is to be zero.

However, that is not the end of the matter. To fix ideas, consider the TT-version of the metric for a plane-polarized wave propagating in the z-direction:

$$d\tau^2 = dt^2 - (1+\varepsilon C)\,dx^2 - (1-\varepsilon C)\,dy^2 - dz^2 \tag{13.26}$$

where for brevity $C=\cos\omega(t-z)$; we shall also use $S=\sin\omega(t-z)$. By the usual method, we may calculate the Riemann tensor, working *to second order in* ε for reasons which will soon appear,

$$R^{tx}{}_{tx} = -R^{tx}{}_{zx} = R^{zx}{}_{tx} = -R^{zx}{}_{zx} = \tfrac{1}{2}\omega^2[\varepsilon C - \varepsilon^2(C^2 - S^2/2)],$$
$$R^{ty}{}_{ty} = -R^{ty}{}_{zy} = R^{zy}{}_{ty} = -R^{zy}{}_{zy} = \tfrac{1}{2}\omega^2[-\varepsilon C - \varepsilon^2(C^2 - S^2/2)].$$

As usual, the other elements are related to these by symmetry, or else are zero. The Einstein tensor is immediate (Problem 10.1), its nonzero elements being

$$G^t{}_t = -G^t{}_z = G^z{}_t = -G^z{}_z = -\varepsilon^2\omega^2(C^2 - S^2/2). \tag{13.27}$$

The terms of first order in ε have now dropped out, as they must since (13.26) is a first-order solution of the Einstein equations. Of course, the presence of the second-order terms is in no way a problem: they will vanish in their turn as soon as the necessary second-order corrections to the solution are incorporated.

However, let us adopt a perverse point of view. Let us insist that the first-order solution is in fact the *exact* solution. Then (13.27) reveals that what ought to be empty space is filled with a spurious matter tensor $\mathbf{T}=\mathbf{G}/8\pi$. To restore the vacuum, we shall have to assign a compensating energy tensor $-\mathbf{T}$ to the gravity wave itself. And because this procedure allows us to keep an undistorted Minkowski space as the backcloth, energy conservation is rescued with the help of this arbitrary introduction of the compensating energy tensor.

The foregoing may not carry much conviction; after all, we are trying to get something nonzero from something that is essentially zero. In any case, the idea collapses for fields which are not weak. On the other hand, for weak fields the introduction of the compensating energy tensor is exactly what we need to be able to talk about a conserved energy; this simplifies certain kinds of calculation substantially.

The energy-current-density $-G^z{}_t/8\pi$ is not constant in time (though the corresponding quantity for a *circularly* polarized wave is constant); we take the time average over one period, $\varepsilon^2\omega^2/4$, as representative of the energy carried by the wave. On the other hand, from (13.26),

$$H^{\mathrm{TT}}_{xx} = -H^{\mathrm{TT}}_{yy} = -\varepsilon C$$

from which it follows that two time averages, $\langle\ldots\rangle$, are equal:

$$-(\omega^2/2)\langle H^{\mathrm{TT}}_{\mu\nu} H^{\mathrm{TT}\mu\nu}\rangle = 8\pi\langle\text{energy-current-density}\rangle. \tag{13.28}$$

Though we have discussed this result for only one polarization, it happens to be true for all.

At large r the radiation from a source at the origin is nearly plane, and we may use this result to evaluate the integrated energy flow (the *radiated power*) over the surface of a large sphere. This is then interpreted as the energy loss at the source.

The symmetric inertia tensor has six independent elements. The traceless part $\mu^-_{ij} \equiv \mu_{ij} - \frac{1}{3}\delta_{ij}$ trace μ therefore has five only; these five may be regarded as the amplitudes of the five distinct radiation processes that may take place. For the radiation in any chosen direction, the TT part is defined by just two amplitudes; we may say that only two processes are effective for that direction. Averaging over all directions gives

$$\text{averaged } \Sigma_{ij}(\mu^{\mathrm{TT}}_{ij})^2 = \tfrac{2}{5}\Sigma_{ij}(\mu^-_{ij})^2. \tag{13.29}$$

The left side has the form we need to evaluate the radiated power; the right side includes the required factor 2/5, and is the only isotropic quadratic combination of the elements of μ^-. This argument relies on the isotropy of 3D space and, though seemingly very condensed, is watertight. A more explicit discussion is given in problem 5.

Putting together (13.28), (13.21) and (13.29) and integrating over a large sphere of radius r immediately gives

$$\text{total radiated power} = \tfrac{1}{5}\omega^6\Sigma_{ij}(\mu^-_{ij})^2,$$

time averaged if necessary.

13.11 The energy radiated by a binary star system

Two stars are in orbit round each other; for simplicity, suppose that the masses are both M and that the orbits are circular, each of radius R and both in the xy-plane. The angular velocity of each star in its orbit is Ω, so that $\phi = \Omega t$. The inertia tensor in this case may be evaluated as

$$2MR^2\begin{bmatrix} \cos^2\Omega t & \cos\Omega t\sin\Omega t & \cdot \\ \cos\Omega t\sin\Omega t & \sin^2\Omega t & \cdot \\ \cdot & \cdot & \cdot \end{bmatrix}.$$

This expression has a constant part, and a part which oscillates with frequency $\omega = 2\Omega$. The constant part is irrelevant for radiation; the remainder is

$$MR^2\begin{bmatrix} \cos\omega t & \sin\omega t & \cdot \\ \sin\omega t & -\cos\omega t & \cdot \\ \cdot & \cdot & \cdot \end{bmatrix}.$$

As this happens to be traceless already, the radiated power is immediate:

$$\begin{aligned}\text{total radiated power} &= \tfrac{1}{5}\omega^6 M^2R^4\,[2\cos^2\omega t + 2\sin^2\omega t] \\ &= \tfrac{128}{5}\Omega^6 M^2 R^4.\end{aligned}$$

In recent years, some very tight, very massive binary systems have been discovered; by ordinary standards, the values of Ω are very large. It is therefore to be expected that the radiated power is substantial. There is of course no possibility of detecting the radiation directly in a terrestrial laboratory. However, energy is radiated at the expense of energy of the angular motion, and a change in angular velocity ought to be observable. In fact, changes in $1/\Omega$ of the order of 10^{-4} s per year have been observed; this is close to the theoretical prediction.

13.12 Angular momentum as a source of gravity

Up to now, only the far field of a compact source has been discussed. This goes like $1/r$, and depends on the mass and the inertia tensor of the source. Other attributes of the source may be expected to generate a field which falls off more rapidly than $1/r$; the most important of these is *angular momentum.*

We return to (13.16), and make an approximation which is less drastic than that of (13.18). We shall consider a mass tensor $\mathbf{T}(\mathbf{r}')$ which is *constant* in time—the most important case—and use the expansion

$$\frac{1}{|\mathbf{r}-\mathbf{r}'|}=\frac{1}{r}+\frac{\mathbf{r}}{r^3}\cdot\mathbf{r}'+\cdots\quad,$$

correct to first order in $\mathbf{r}'$. To this order, (13.16) becomes

$$H_{\mu\nu}(\mathbf{r})=\frac{1}{r}K_{\mu\nu}+\frac{\mathbf{r}}{r^3}\cdot\mathbf{L}_{\mu\nu}+\cdots\quad,$$

in which $\mathbf{K}$ is as before, and $\mathbf{L}$ is the **first moment** of the matter tensor,

$$\mathbf{L}_{\mu\nu}=-4\int_{\text{cross-section}}\mathbf{r}'T_{\mu\nu}(\mathbf{r}')\,\mathrm{d}^3\mathbf{r}'.$$

Thus the first moment is the source of a field which falls off as $1/r^2$.

The most interesting first moment arises from an unchanging distribution of momentum-density (an example is provided by a rigid uniform sphere rotating with constant angular velocity). The six nonzero elements of the matter tensor are

$T_{ti}=T_{it}=-p_i$, the i-component of the momentum-density, for $i=x,y,z$.

Consequently, $\mathbf{L}$ involves the nine integrals

$$J_{ij}=2\int r_ip_j\,\mathrm{d}^3\mathbf{r},\quad \text{for } i \text{ and } j=x,y,z.$$

Now, $\mathbf{p}$ is unchanging, and the continuity equation is $\nabla\cdot\mathbf{p}=0$; applying Green's theorem to the divergence discussed in section 13.7 shows that $\mathbf{J}$ is antisymmetric:

$$J_{ij}=-J_{ji}=\int(r_ip_j-r_jp_i)\,\mathrm{d}^3\mathbf{r}.$$

In fact, $\mathbf{J}$ is the **total angular momentum** of the source: see equation (10.11).

We shall ignore the effect of $\mathbf{K}$ (it may be added in later, if desired). Then the nonzero elements of **H** are

$$H_{ti} = H_{it} = \frac{2}{r^3} r_j J_{ji}. \tag{13.30}$$

Since in this case $H^{\mu}{}_{\mu} = 0$, these are equally the values of h_{ti} and h_{it}. At this point it is illuminating to go over into a 3D vector notation in which we use the vectors

$$\mathbf{h} = (h_{tx}, h_{ty}, h_{tz}), \quad \mathbf{r} = (x, y, z), \quad \text{and} \quad \mathbf{J} = (J_{yz}, J_{zx}, J_{xy}).$$

Then (13.30) may be rewritten

$$\mathbf{h} = (2/r^3)\mathbf{J} \wedge \mathbf{r},$$

a neat way of describing the gravity field produced by the angular momentum **J**.

13.13 The Lense–Thirring effect

The Christoffel symbols of interest for this section are

$$\Gamma^{i}{}_{tj} = -\Gamma_{itj} = -\tfrac{1}{2}(\partial_j h_{ti} - \partial_i h_{tj}), \quad i \text{ and } j = x, y, z.$$

(There are others which are nonzero, but they are insignificant in what now follows.) Let us introduce a further 3D vector

$$\boldsymbol{\Omega} = (\Gamma^{y}{}_{tz}, \Gamma^{z}{}_{tx}, \Gamma^{x}{}_{ty}) \equiv \tfrac{1}{2}\nabla \wedge \mathbf{h}.$$

We shall see that $\boldsymbol{\Omega}$ describes the more important gravitational influences exerted on the environment by an angular momentum.

First consider the *slow* motion of a particle along a given track $\mathbf{r} = \mathbf{r}(t)$; the four-vector **F** is

$$F^t = \text{nearly zero},$$
$$F^x = \ddot{x} + \Gamma^{x}{}_{ty}\dot{t}\dot{y} + \Gamma^{x}{}_{tz}\dot{t}\dot{z}. \text{ etc.}$$

In terms of the 3D vector $\mathbf{f} = (F^x, F^y, F^z)$, using $\dot{t} = 1$ nearly, these equations may be written in the 3D form

$$\mathbf{f} = \ddot{\mathbf{r}} - \boldsymbol{\Omega} \wedge \dot{\mathbf{r}}.$$

The second term is an extra velocity-dependent acceleration reminiscent of the negative of a Coriolis force. ('Negative', since **f** is the *real* force per unit mass required to keep the particle on track against such apparent forces.) On account of the rotation of the Earth, for example, a particle in free radial fall at the Equator is deflected towards the east, while a particle passing across the North Pole will be deflected to the 'left' in a natural sense. At other latitudes the effect is not so simply described, on account of the involved way in which $\boldsymbol{\Omega}$ depends on position. The phenomenon is picturesquely (though improperly) described as **dragging**. It is far too insignificant to be observable by any terrestrial or astronomical technique currently available.

The effect on parallel transport is of interest. An inertial observer hovering at a fixed point experiences a rotation relative to the coordinate system. Since his velocity vector is $(1, 0, 0, 0)$, only the matrix Γ_t is relevant to the parallel transport equations

(5.8); in fact, the equations for the 3D part of the vector are

$$\frac{\mathrm{d}}{\mathrm{d}t}\begin{bmatrix}\lambda_x\\ \lambda_y\\ \lambda_z\end{bmatrix} = -\begin{bmatrix}\cdot & \Omega_z & -\Omega_y\\ -\Omega_z & \cdot & \Omega_x\\ \Omega_y & -\Omega_x & \cdot\end{bmatrix}\cdot\begin{bmatrix}\lambda_x\\ \lambda_y\\ \lambda_z\end{bmatrix}.$$

In 3D vector form, this may be written

$$\dot{\boldsymbol{\lambda}} = \boldsymbol{\Omega} \wedge \boldsymbol{\lambda},$$

which declares that the vector $\boldsymbol{\lambda}$ 'precesses' about the direction $\boldsymbol{\Omega}$ in the positive sense with angular velocity $|\boldsymbol{\Omega}|$. This is the **Lense–Thirring effect**.

There is some hope that the effect may be just observable by current techniques, by monitoring a gyroscope mounted in an Earth satellite in a strictly polar orbit. (For any other orbit, the effect will be overwhelmed by the geodesic effect; see section 5.14.) The sign of the effect changes four times per orbit, but the aggregate effect is not zero and is therefore cumulative.

13—Notes and problems

1. Verify that with neglect of second-order terms

$$(\gamma^{\mu\rho} - h^{\mu\rho})(\gamma_{\rho\nu} + h_{\rho\nu}) = \delta^{\mu}{}_{\nu}.$$

2. In section 13.4, note that the transition from one form of the gauge condition to the other is not quite straightforward: the terms with $\exp \mathrm{i}k_\mu x^\mu$ need to be treated separately from those with $\exp(-\mathrm{i}k_\mu x^\mu)$. It is a matter of keeping the sign right.

3. For the purposes, the **Dirac function** $\delta(x)$ is a notional function, devised to pick out a particular point in an integration to the exclusion of others:

$$\int_{-\infty}^{\infty} \delta(x-y)f(y)\,\mathrm{d}y \equiv f(x).$$

By using the appropriate change of variable $z = z(y)$ in this integral, show that

$$\delta(z - z_0)\frac{\mathrm{d}z_0}{\mathrm{d}y_0} = \delta(y - y_0),$$

where $z_0 = z(y_0)$. This relation was used in the final step to (13.16). The results on the inhomogeneous wave-equation given in section 13.5 are standard, and are readily available in the literature.

4. Show that in nD, the reverse of the relation

$$H^{\rho}{}_{\sigma} \equiv h^{\rho}{}_{\sigma} - \tfrac{1}{2}\gamma^{\rho}{}_{\sigma}h^{\alpha}{}_{\alpha}$$

is

$$h^{\rho}{}_{\sigma} \equiv H^{\rho}{}_{\sigma} - \frac{1}{n-2}\gamma^{\rho}{}_{\sigma}H^{\alpha}{}_{\alpha}.$$

(To begin, set $\rho = \sigma$ in the first relation and sum, to get $H^{\alpha}{}_{\alpha}$ in terms of $h^{\alpha}{}_{\alpha}$.) For the case $n = 4$ which interests us, the relations for $H \to h$ and for $h \to H$ take the same form.

5. (i) Show from the recipe of (13.23) that

$$\mu_{ij}^{\mathrm{TT}}\mu_{ij}^{\mathrm{TT}} = \mu_{\bar{ij}}\mu_{\bar{ij}} - 2u_i u_j \mu_{\bar{ik}}\mu_{\bar{kj}} + \tfrac{1}{2}u_i u_j u_k u_l \mu_{\bar{ij}}\mu_{\bar{kl}}\,.$$

(ii) Show that, if **a** is a constant vector, the average of $(\mathbf{a}\cdot\mathbf{n})^{2p}$ over all directions **n** is

$$\frac{1}{4\pi}\int (a_i n_i)^{2p}\,\mathrm{d}^2\Omega = \frac{({a_1}^2 + {a_2}^2 + {a_3}^2)^p}{2p+1}.$$

(Write $\mathbf{a}\cdot\mathbf{n} = |a|\cos\theta$ in suitably chosen spherical polar coordinates.)

(iii) Consider the cases $p = 1, 2$ to obtain

$$\text{averaged } n_i n_j = \tfrac{1}{3}\delta_{ij},$$

and

$$\text{averaged } n_i n_j n_k n_l = \tfrac{1}{15}(\delta_{ij}\delta_{kl} + \delta_{ik}\delta_{jl} + \delta_{il}\delta_{jk}).$$

(In the second case, due regard has to be paid to the necessary symmetries.)

(iv) Use the results of (i) and (iii) to obtain (13.29).

6. Two point masses m, M are separated by a distance d. Show that the total moment of inertia about the centre of mass is $m_{\mathrm{R}}d^2$, where the **reduced mass** $m_{\mathrm{R}} = mM/(m+M)$.

The two masses revolve about their centre of mass; the orbit of each is circular. The period of one revolution is T. Show that the total radiated gravitational power is $(4\pi/T)^6 {m_{\mathrm{R}}}^2 d^4/10$.

Show that Jupiter in its orbit round the Sun radiates about 5.2 kW. (For Jupiter: mass = 1.42 m, distance from Sun = 7.78×10^{11} m, 'year' = 11.9 Earth-years, to be converted to metres. See section 10.12 for the conversion to watts.)

Jupiter radiates the most. Venus comes next, with about 0.66 kW. The Solar System as a whole cannot radiate much more than about 6 kW of gravity wave energy.

7. In the general case of a static mass distribution, not necessarily compact, H_{tt} is effectively the Newtonian gravitational potential $\phi(\mathbf{r})$ of the distribution:

$$H_{tt} = 4\phi(\mathbf{r})$$

Obtain the *Newtonian approximation*

$$\mathrm{d}s^2 = (c^2 + 2\phi(\mathbf{r}))\,\mathrm{d}t^2 - (1 - 2c^{-2}\phi(\mathbf{r}))(\mathrm{d}x^2 + \mathrm{d}y^2 + \mathrm{d}z^2)$$

in which c has been reinserted for the sake of the units. (This gives a generalization of the approximate Schwarzschild metric of section 13.6.)

Retrospect

Einstein's Theories of Relativity are theories about physics. They have led to much that is exciting in mathematics, and we should be grateful for that. However, mathematical fertility is not the same as physical truth, and a fundamental question is: How far can we rely on General Relativity as an adequate description of the part of physics that it claims to talk about?

In practical terms, the results are impressively good. Never mind that the differences between relativity gravity and Newtonian gravity are tiny: the point is that observation shows that the differences are there, and that therefore the Newtonian description is at fault in some respect. General Relativity—when we know how to apply it—gets it right in every case, to within observational error. It is therefore a good candidate for consideration.

Moreover, General Relativity conforms to the general trend of the last century or so: instantaneous action at a distance has been replaced by effects propagating according to some wave equation or other. The first clear example—apart from acoustic waves and the like—was Maxwell's equations for the propagation of electromagnetic effects. (The Laplace and Poisson equations don't really count; they are nothing more than reformulations of the inverse-square law.) The last citadel of instantaneous action at a distance has been the backcloth itself, the flat space of pre-relativity physics: its very rigidity bears witness to this. Spacetime is no longer rigid, according to Einstein; its behaviour is controlled by field equations, just like everything else.

Will the Einstein theory ever be bettered? It is essential that it is. The twentieth century has seen the arrival of *two* great restructurings: Relativity and Quantum Theory. It is fair to say that in every case where we have been able to apply the one or the other the results have been superlatively good. It is therefore very strange that reconciling the two—let alone merging them into a single theoretical structure—should present such intractable difficulty. Yet it does. Why should two extraordinarily successful descriptions be at such odds? No-one is sure, and it looks as if we shall remain in the dark for a very long time, in spite of the frenzy of labour that is going on.

The Universe is a strange place, and it would be wrong to think that we understand it just because we have been shown how to describe a minuscule aspect of it so well.

Honour where honour is due: a word must be said about James Clerk Maxwell (1831–79). The final form of the equations for electromagnetism was developed by him, and was a crucial component in Einstein's search for a proper theory. Special Relativity was not to appear for another forty years, and yet Maxwell's equations satisfy all the requirements of the theory: this is why the propagation of *electromagnetic* radiation plays such a large part in the development (see, for example, Chapter 2). Maxwell got it right.

At the age of seventeen, Maxwell wrote:

> Of forces acting between two particles of matter there are several kinds.
>
> The first kind is independent of the quality of the particles, and depends solely on their masses and mutual distance. Of this kind is the attraction of gravitation...
>
> The second kind... are the attractions of magnetism [and] electricity...

So Maxwell seems to have understood clearly that gravity is *kinematic*. When he was about thirty, he published his celebrated equations. At forty-eight he died. If he had survived to the end of the century, would it have been Einstein's name that we remember now? I wonder.

Index